The Century World's Fair Book

for Boys and

Being the Adventures of Harry a

Tutor, Mr. Douglass, at the W

Exposition

Tudor Jenks

Alpha Editions

This edition published in 2024

ISBN : 9789366389202

Design and Setting By
Alpha Editions
www.alphaedis.com
Email - info@alphaedis.com

Contents

THE WHITE CITY.

VIEW FROM THE FERRIS WHEEL.

THE ADMINISTRATION BUILDING.

CHAPTER I

Started by Cable — The Journey by Sleeper — Arrival in Chicago — Finding Rooms — The Fair at Last!

HERE ARE THE
TICKETS.

"MR. DOUGLASS wants to see you, Master Harry," said the maid, coming to the door of the boys' room.

"What's he found out now, I wonder?" said Harry to Philip, in a low tone. "I don't remember anything I have done lately."

"He's in a hurry, too," said the girl, closing the door.

Harry ran down to Mr. Douglass's room on the first floor. The two boys were beginning their preparation for college, and were living in a suburb of New York city with their tutor, Mr. Douglass, a college graduate, and a man of about thirty-five. Harry's father, Mr. Blake, was abroad on railroad business, and did not expect to return for some months. Philip was Harry's cousin, but the two boys were very unlike in disposition—as will be seen. Their bringing up may have been responsible for some of the differences in traits and character, for Harry was a city boy, while his cousin was country-bred.

When Harry knocked at the door of Mr. Douglass's study, he knew by the tutor's tone in inviting him in that the teacher had not called him simply for a trivial reprimand. It was certainly something serious; perhaps news from Harry's father and mother.

"Sit down, Harry," said the tutor,—"and don't be worried," he added, seeing how solemn the boy looked. "I have had a message by cable from your father; but it's good news, not bad. Read it."

He handed Harry the despatch. It read:

> Take Hal and Phil to Fair. My expense. Letter to Chicago. See Farwell about money and tickets.

"Rather sudden, isn't it?" said Mr. Douglass, smiling.

THE FOUNDATION OF THE MANUFACTURES

"Yes," said Harry, "but—immense! Don't you think so?"

"I'm glad to go," the tutor said. "It seems to me that a visit to the Fair is worth more than all the studying here you boys could do in twice the time you'll spend there; and it's a lucky opportunity for me."

"Then you'll go?" said Harry, to whom the news seemed a bit of fairy story come true, with the Atlantic cable for a magic wand.

"Of course," answered the tutor. "The only thing that surprises me is the quickness of your father's decision."

"That's just like him," said Harry. "He's a railroad man, you know, and they always go at high pressure. Why, he'd rather talk by telephone, even when he can't get anything but a buzz and a squeak on the wire, than send a messenger who'd get there in half the time."

"But has he said anything about sending you before?"

"No. The fact is, people abroad are slow to know what a whacker this Fair is! They think it's a mere foreign exposition. Father's just found out that Uncle Sam has covered himself with glory, and now he wants Phil and me to see the bird from beak to claws—the whole American Eagle."

"But sha'n't we have trouble about tickets?" asked Mr. Douglass.

"No," said Harry. "Father's a railroad man. That's what 'See Farwell' means. You let me go to see him. He's the general manager, or some high-cockalorum. He'll see us through by daylight."

"Very well," said Mr. Douglass, "I'm just as glad to go as you are. Philip and I will attend to the packing, and you shall go to New York this afternoon and see Mr. Farwell. Now you can tell Philip about it."

AND LIBERAL ARTS BUILDING.

Harry ran out of the room, slamming the door behind him, but Mr. Douglass only laughed. Perhaps he would have slammed it, too, if he'd been in the boy's place.

"Well?" said Philip, looking up from the Xenophon he was translating.

"Thanks be to Christopher Columbus!" said Harry, with a jig-step.

"Has he done anything new?" Philip asked, looking over his spectacles.

"I guess not," said Harry, "but we're going to the Fair."

"How can we?" Philip asked.

Harry threw the cable despatch down upon the table, and turned to get his hat. Philip read the telegram, carefully wiped his glasses, rose, put the Xenophon into its place upon his book-shelves, and said:

"Xenophon will have to attend to his own parasangs for a while."

"You pack up for me, and I'll see to the railroad-tickets," said Harry. "I have just about time to catch the train for New York."

That was a hard and busy day for all three of the party. Perhaps Harry's share was the easiest, for, by showing his father's despatch to Mr. Farwell, he had everything made easy for him. Still, even influence might not have secured them places except for the aid of chance. It happened that a prominent man had, at the last moment, to give up a section in the Wagner sleeper, and this was turned over to Harry. So, late in the afternoon the boy came back with what he called "three gilt-edged accordion-pleated tickets."

Meanwhile Mr. Douglass and Philip had put into three traveling-bags as much as six would hold, and the party went to bed early to get a good rest before the long journey.

A WILDERNESS OF IRON.—BUILDING MACHINERY HALL.

Next day at nearly half-past four the three travelers walked through the passageway at the Grand Central Depot, had their tickets punched,—and Philip noticed that the man at the gate kept tally on a printed list of the numbers of different tickets presented,—and entered the mahogany and blue-plush Wagner cars.

In a few minutes some one said quietly: "All right," and the train gently moved out.

"I can remember," said Mr. Douglass, "when a train started with a shock like a Japanese earthquake. Now this seemed to glide out as if saying, 'Oh, by the way, I think I'll go to Chicago!'"

Harry laughed. "Yes," he said, "and how little fuss there is about it. Why, abroad, I remember that they had first a bell, then a yell, then a scream, then the steam!"

As the train passed through the long tunnel just after leaving the station, Mr. Douglass remarked:

"How monotonous those dark arches of brickwork are!"

"Yes," said Philip, "they should have a set of frescos put in them."

"But no one could see the pictures," said Mr. Douglass, "we pass them so fast."

"That's true," said Harry, with a pretended sigh; "but they might have to be instantaneous photographs."

Philip looked puzzled for a minute and then laughed. After they left the tunnel, they passed through the suburbs of New York, entered a narrow cut that turned westward, and were soon sailing along the Hudson River—or so it seemed. There was no shore visible beside them, except for an occasional tumble-down dock, and beyond lay the river—a soft, gray expanse relieved against the Palisades, and later against more distant purple hills. It was a rest for their eyes to see only an occasional sloop breaking the long stretch of water, and the noise of the train was lessened because there was nothing to echo back the sounds from the river.

Mr. Douglass found his pleasure in the scenery, the widenings of the river, the soft outlines of the hills, the long reflection of the setting sun. But the boys cared more to see the passengers.

"Isn't it funny," said Philip, "how Americans take things as a matter of course? I really believe that if the train was a sort of Jules Verne unlimited express for the planet Mars, the people would all look placid and read the evening papers."

"Of course," said Harry. "What else can they do? Would you expect me to go forward and say: 'Dear Mr. Engineer, but *do* you really think you know what all these brass and steel things are? Don't you feel scared? Won't you lie down awhile on the coal, while I run the engine for you?'"

"Nonsense!" said Philip, laughing. "But they might show some interest."

"They do," said Harry; "but that's not what I'm thinking of. I'm thinking I'll be a civil engineer."

"Why?" said Philip.

"Just think," Harry answered, pointing from the car window, "what a good time they must have had laying out this road! Why, it was just a camping-out frolic, that's all it was."

ONE OF THE DECORATORS AT WORK.

"Didn't you hear the waiter say dinner was ready?" said Mr. Douglass.

"No," said Philip; "but I knew it ought to be, if they care for the feelings of their passengers. Where is the dining-car?"

"At the end of the train," said Mr. Douglass. "Come, we'll walk through."

So, in single file ("like cannibals on the trail of a missionary," Harry said), they passed from car to car. The cars were connected by vestibules—collapsible passageways, folding like an accordion—and it was not necessary to go outside at all. The train was an unbroken hallway.

"It is much like a long, narrow New York flat," said Philip. "People who live in flats must feel perfectly at home when they travel in these cars."

They found the dining-car very pretty and comfortable. Along one side were tables where two could sit, face to face. On the opposite side of the aisle the tables accommodated four. The boys and their tutor took one of the larger tables. The bill of fare was that of a well-appointed hotel or restaurant,—soup, fish, entrées, joint, and dessert,—and it was difficult to realize that they were eating while covering many miles an hour; in fact, the only circumstance that was a reminder of the journeying was a slight rim around the edge of the table to keep the dishes from traveling too.

MAKING "STAFF."

"It is strange," said Mr. Douglass, "how people have learned to eat dishes in a certain order, such as you see on a bill of fare. Probably this order of eating is the result of tens of millions of experiments, and therefore the best way."

THE "COURT OF HONOR" AS IT LOOKED IN JUNE, 1892.

"The best for us," said Philip; "but how about the Chinese?"

Mr. Douglass had to confess himself the objection well taken.

"I believe the Chinese were created to be the exceptions to all rules," he said.

The dining-car had an easy, swaying motion that was very pleasant, and altogether the dinner was a most welcome change from the ordinary routine of a railway journey.

As the boys walked back to their own section, Philip noticed a little clock set into the woodwork at one end of the smoking-car. He was surprised to see that it had two hour-hands, one red and one black.

He pointed it out to Mr. Douglass, who told him that the clock indicated both New York and Chicago times—which differ by an hour, one following what is called "Eastern," the other "Central" time.

By the time they were again settled in their places it was dark outside; and, as Philip poetically said, they seemed to be "boring a hole through a big dark." One of the colored porters looked curiously at Philip, as if he had overheard this remark without understanding its poetical bearing.

"He thinks you are a Western desperado!" said Harry, with a grin.

"Boys," said Mr. Douglass, "the porters will soon make up the beds, and I want you to see how ingeniously everything is arranged."

Here is what the porter did:

He stood straddling on two seats, turned a handle in the top of a panel, and pulled down the upper berth. It moved on hinges, and was supported after the manner of a book-shelf by two chains that ran on spring pulleys.

Then he fastened two strong wire ropes from the upper to the lower berths.

"What's that for?" asked Harry.

"To prevent passengers from being smashed flat by the shutting up of the berth," Philip answered, after a moment's puzzling over the question.

"You can have the upper berth, Philip," said Harry, impressively. "It's better ventilated than the lower, they say; but I don't mind that."

Meanwhile the porter took from the upper berth two pieces of mahogany, cut to almost fill the space between the tops of the seats and the side roofs of the car. The edges were grooved, and slid along upon and closely fitted the top of the seat and a molding on the roof. These side-pieces were next fastened by a brass bolt pushed up from the end of the seat-back.

Then the bed-clothing (kept by day in the lower seats and behind the upper panel) was spread on the upper berth, and the mattress of the lower berth was made up from the seat-cushions, supported upon short slats set from seat to seat.

While the beds were being made, the boys were amused to see some ladies laughing at the man's method of getting the clothes and pillows into place. A woman seems to coax the bed into shape, but a man bullies it into submission.

"They think it's funny to see him make a bed," said Harry, in an undertone; "but if they were to try to throw a stone, or bait a fish-hook, I guess the darky would have a right to smile some too."

To finish his work, the porter hung a thick pair of curtains on hooks along a horizontal pole, and then affixed a long plush strip to which were fastened large gilt figures four inches high—the number of the section.

"It would be fun to change the numbers around," remarked Harry, pensively. "Then nobody would know who he was when he got up. But perhaps it would make a boy unpopular if he was caught at it."

"YO' SECTION
READY, SAH!"

Mr. Douglass admitted that it might.

As the porter made up their own section, Harry pulled out his sketch-book and made a little picture of him.

"It's hard times on the railroad now," he remarked, as he finished the sketch. "See how short they have to make the porters' jackets! But it must save starch!"

The boys had wondered how the people would get to bed, but there seemed no difficulty about it. As for our boys, who had the upper berth, one by one they took off their shoes, coats and vests, etc., and then climbed behind the curtains, where they put their pajamas over their underclothes.

After they were in bed, they talked but little, for they were tired.

"This rocking makes me drowsy," Philip said; "it's like a cradle."

“Yes,” Harry answered, as the car lurched a little—“a cradle rocked by a mother with the St. Vitus’s dance!”

While going to sleep, the boys were puzzled to account for the strange noises made by the train. At times it seemed to have run over a china-shop, and at other times the train rumbled hoarsely, as if it were running over the top of an enormous bass-drum.

Soon the great train was transporting two boys who were fast asleep in Section No. 12; they woke fitfully during the night, but only vaguely remembered where they were, until the cold light of morning was reflected from the top of the car.

Dressing was more difficult than going to bed, but by a combination of patience and gymnastics Harry and Philip were soon able to take places in the line that led to the wash-room. Thence, later, they came forth ready for breakfast (for which they had to “line up” again), and another all-day ride.

THE OPENING OF THE FAIR, MAY 1, 1893.—THE PRESIDENT OF THE UNITED STATES SPEAKING.

At breakfast, the next table to them was occupied by a gentleman named Phinney, and his son. Harry knew the son slightly, having once been his schoolmate. Young Phinney was making a second visit to the Fair, and he told Harry that on the former trip the train had run around Niagara Falls in such a way as to give the passengers an opportunity to view them.

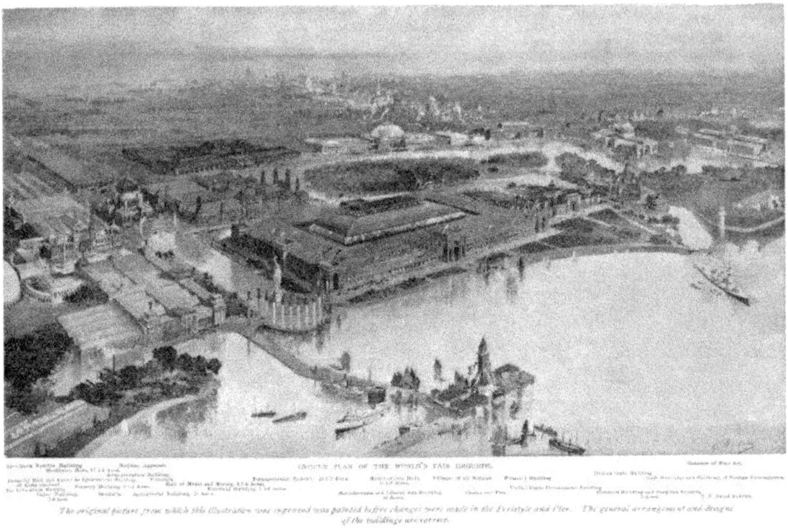

GROUND PLAN OF THE WORLD'S FAIR GROUNDS.

Live-Stock Exhibit Building.
Railway Approach.
Galleries of Fine Art.
Machinery Hall, 1-2 Acres.
Administration Building.
Illinois State Building.
Assembly Hall and Annex to Agricultural Building.
Fountain.
Transportation Exhibit, 18 2-3 Acres.
Horticultural Hall, 6 1-2 Acres.
Villages of all Nations.
Women's Building.
State Buildings and Buildings of Foreign Governments.
63 Acres reserved for Live-Stock Exhibit.
Hall of Mines and Mining, 8 3-4 Acres.
Forestry Building, 2 1-2 Acres.
Electrical Building, 9 3-4 Acres.
United States Government Building.
Dairy Building, 3-4 Acre.
Sawmills.
Agricultural Building, 15 Acres.
Manufactures and Liberal Arts Building, 44 Acres.
Casino and Pier.
Fisheries Building and Deep-Sea Aquaria, 3 Acres.
U. S. Naval Exhibit.

The original picture from which this illustration was engraved was painted before changes were made in the Peristyle and Pier. The general arrangement and designs of the buildings are correct.

The train had stopped there for five minutes, and they had climbed down near the rapids to a point where there was an excellent view of "the great cataract"—so young Phinney called it. He gave the boys some pictures showing the falls, and indeed there was a picture of the falls upon the side of the breakfast bill of fare.

During the forenoon the train was passing through Canada—the boys' impression of that country being a succession of flat fields, ragged woods, sheep, swine, and a few pretty, long-tailed ponies grazing upon browning turf. Philip said that it was like "the Adirondacks spread flat by a giantess's rolling-pin."

At Windsor the train, separated into sections, was run upon a ferry-boat (upon which one small room was marked "U. S. Customs") and carried over to Detroit. Here Mr. Douglass made the boys laugh by suddenly jumping back from the window. He had been startled by a large round brush that was poked against the window from outside to dust it.

From Detroit the train ran through Michigan—mainly through a flat country of rich farming land. Philip, who had never been West, was much surprised at the uninterrupted stretches of level ground. Mr. Douglass asked him what he thought of the region. Philip adjusted his glasses and replied slowly: "Well, it's fine for the farmers, but it is no place for speaking William Tell's piece about 'Ye crags and peaks, I'm with you once again!'"

"You must not forget, though," said Mr. Douglass, "that it is the rich farming lands that really underlie America's prosperity. When you see the Fair, you will understand better what a rich nation we are; but without our great wheat-lands we should, like England, be dependent upon commerce for our very existence."

The boys were much less talkative as the train neared Chicago. They were somewhat tired, and were also thinking of the amount of walking and sight-seeing that was before them.

All at once, at about half-past five, New York time (for the travelers had not yet changed their watches to an hour earlier), Mr. Douglass pointed out of the right-hand forward window. Both boys looked. There, in the distance, rose above the city houses a gilded dome, and from the opposite car-window they saw just afterward a spider-web structure.

"I know it!" Philip sang out; "that's the Administration Building. But what is the other?"

"The Ferris Wheel," answered Harry.

"Yes," said the tutor, "we are going to leave the car not far from the Plaisance gate."

"Sixtieth street next!" cried the brakeman.

"Come, we get out here. It's nearest the grounds, and I have been told it is wise to lodge as near as possible."

When the cars stopped, the party descended upon a platform with "rails to the right of them, rails to the left of them," and trains and crowds in all directions. Mr. Douglass led the way out into the huddled settlement of apartment-houses, hotels, and lodgings that has sprung into existence around Jackson Park, the Fair Grounds.

Then began their search for rooms. At first it seemed discouraging; neatness outside was not always a sign of what to expect inside. They labored up-stairs and down again several times. At one attractive private house they entered, expecting quiet, homelike rooms. In the tiny parlor they found five cots set "cheek by jowl" as close as they could be jammed. They smiled at this, but found the rest of the rooms as fully utilized. Mr. Douglass made some objection, and was told by the self-possessed landlady that "some very fine gentlemen thought her fifty-cent beds were very elegant." At another house they were passing, a boy who couldn't have been over five years old rushed out like a little Indian on the warpath, crying, "Hi! You lookin' fer rooms?" Amused at the little fellow's enterprise, our travelers followed him, the boy going forward on his sturdy little legs, and crying, "Hi, there, Mama! Here's roomers! I got you some roomers!"

"HI, THERE, MAMA!
HERE'S ROOMERS!"

But unfortunately the boy proved more attractive than the rooms. After a long walk, but without going far from the Fair Grounds, they took rooms at a very good hotel. The price was high, perhaps, but reasonable considering the advantages and the demand for lodgings. They took two rooms, one with a double bed for the boys, the other a single room for the tutor.

Gladly they dropped the satchels that had made their muscles ache, and after leaving the keys of their rooms with the hotel clerk, they set forth for their first visit to the Fair. In order that guests should not forget to leave their keys, each was inserted at right angles into a nickel-plated strip of metal far too long to go comfortably into the pocket even of an absent-minded German professor.

"One advantage of being in a hotel," said Mr. Douglass, as they walked toward the entrance of the grounds, "is the fact that on rainy, disagreeable days we can get meals there if we choose. It is not always pleasant to have to hunt breakfast through the rain. But usually we shall dine where we happen to be in the grounds; there are restaurants of all sorts near the exhibits, from a lunch-counter up."

Along the sidewalk that led from their hotel to the entrance were dining-rooms, street-peddlers' counters, peddlers with trays—all meant as inducements to leave money in the great Western metropolis. One thing the boys found very amusing was an Italian bootblack's stand surrounded on three sides by a blue mosquito-netting.

"If it had been on all sides," said Harry, "I could have understood it, because it might be a fly-discourager. But now I think it must be only a way of attracting attention."

They had arrived, luckily, on a "fête night." Though tired and hungry, they all agreed that it would never do not to take advantage of so excellent a chance to secure a favorable first impression. So they bought tickets at a little wooden booth, and, entering a turnstile one by one, were at last in the great White City.

HERE WE ARE!

ADMINISTRATION BUILDING.

CHAPTER II

The Fête Night — Rainbow Fountains — The Search-lights — On the Lake — The Fireworks — Passing a Wreck — Diving in the Grand Basin.

"WELL," remarked Harry, as the wicket turned and let him into the grounds, "if any one wishes to take down what I said on entering the grounds, he can write down these thrilling words: 'Here we are at last!'"

"We won't try to do more than get a general idea of things to-night," said Mr. Douglass. "We shall find claims upon our eyesight at every step. But what a crowd!"

The crowd was certainly enormous. At first most of the people seemed to be coming out, but this idea was a mistake. It came from the fact that those going the same way as our party attracted their attention less than those whom they met and had to pass.

A TICKET OF ADMISSION.

They walked between the Pennsylvania Railroad exhibit and the Transportation Building, and entered the Administration Building, which seemed the natural gateway to the Court of Honor and its Basin—always the central point of interest. The paving seemed to be a composition not unlike the "staff" that furnished the material for the great buildings, the balustrades, the statues, and the fountains. It was just at dusk, and the light was soft and pleasant to the eyes. Once in the Administration Building, all our sight-seers threw back their heads and gazed up within the dim and distant dome enriched by its beautiful frescos.

INTERIOR OF THE DOME OF THE ADMINISTRATION BUILDING.

"I have heard," said the tutor, who felt bound to serve as guide so far as his experience would warrant, "that people are unable to understand the vastness of St. Peter's dome at Rome. This dome is even higher, and so I feel sure that, large as it seems to us, our ideas of it fall far below the reality. However, we shall see this many times. Let us go on through, and see the Court of Honor."

Leaving by the east portal, the three came out upon the broad plaza that fronts the basin. By this time the sky was a deep, dark blue, and every outline of the superb group of buildings was sharply relieved.

For a while the three stood silent. There was nothing to say; but each of them felt that the work of men's hands—of the human imagination—had never come so near to rivaling Nature's inimitable glories. The full moon stood high above the buildings at their right, but even her serenity could not make the great White City seem petty.

The boys knew no words to express what they felt. They only knew that in their lives they had never been so impressed except when gazing upon a glorious sunset, an awe-inspiring thunderstorm, or the unmeasured expanse of the ocean.

Philip was the first to speak.

"*Must* it be taken down? *Why* couldn't they leave it? It is—unearthly!"

"Boys," said Mr. Douglass, "I don't preach to you often, and certainly there is no need of it now. But, at one time or another, each of us has tried to imagine what Heaven could be like. When we see *this*," and he looked reverently about him, "and remember that this is man's work, we can see how incapable we are of rising to a conception of what Heaven might be."

A GROUP OF STATUARY ON THE ADMINISTRATION BUILDING, "THE GLORIFICATION OF WAR."

But their rhapsodies could not last long in such a pushing and thronging time. People brushed against them, talking and laughing; the rolling-chairs zigzagged in and out, finding passageway where none appeared; distant bands were playing, and all about them was the living murmur of humanity. Groups were sitting upon every available space: tired mothers with children, young men chatting, and serious-faced country people plodded silently along amid their gayer neighbors.

THE GREAT FOUNTAIN, "THE TRIUMPH OF THE REPUBLIC." Designed and modeled by Frederick MacMonnies. This picture drawn by Mrs. MacMonnies.

For a time the three wandered almost without purpose; then, reaching the further end of the Basin, they looked back at the superb MacMonnies Fountain—the galley that bore the proudly poised figure of Progress.

Opposite, and facing the fountain, rose the massive but perhaps less expressive statue of the Republic. Though the boys were speechless with admiration, delight, and wonder, they found—as others have done—that fine sights do not satisfy the appetite any better than fine words butter parsnips. So Harry turned to Mr. Douglass, saying, "Mr. Douglass, don't you hear the dinner-horn? It seems to me that I do."

A NEARER VIEW OF THE FOUNTAIN.

"All right," he answered; "let us go over to the Casino restaurant and have a comfortable dinner; but first suppose we stop a moment for a look into the Electricity Building. I saw by a program posted up near the entrance that it is open to-night."

As they came nearer, they found the crowd rapidly increasing in density; and when they entered, passing the heroic statue of Franklin, they found themselves entirely at the mercy of the moving throng of people. So thick were the sight-seers packed that the boys could see little except the great Edison Pillar, and that was visible only because it rose so high in air. While they watched the pillar, incrusted with incandescent lights, different colored bulbs sprang into glowing life or faded out, showing a kaleidoscope of patterns changing continually.

"We sha'n't get any dinner if we don't get out now," said Philip, who was struggling to keep his eye-glasses from being displaced.

"Come, then," said Harry; and they turned to stem the tide. For a time they made slight progress; but, luckily, a row of wheeling-chairs came charging slowly but firmly, cutting a path by gentle persistence. Falling in behind these pioneers, they succeeded in escaping to the open air, and then made their way to the Casino. Just before reaching this great restaurant, they saw the convent of La Rábida, which appeared between the Agricultural Building and the Casino.

"See!" said Philip. "There's the model of the convent. Do you know what it reminds me of? It is like a little gray nun sitting demurely in the corner of a grand ball-room!"

And, indeed, the unpretending little building was a distinct rest to the eye, after the proud proportions of its surroundings. As the statues spoke of the future, the convent reminded one of the past.

Entering the Casino brought them back sharply to the present, with its needs and its inconveniences. The prosaic need for dinner was the first to be thought of, and, enormous as was the restaurant, the crowd that night filled every seat, and left plenty of stragglers to stand watchfully about, eager to fill themselves and any vacant chair.

"Boys," said the tutor, sadly, "if we stand here an hour, it will be only a piece of luck if we find a place. Where shall we go?"

"I heard a man say that there was a lunch-counter in the southeastern corner of the Manufactures, etc., etc., Building," said Harry. "This is no time for French bills of fare and finger-bowls. Come, let's go over there."

No one cared to argue the question, and, keeping the lake on their right, they crossed to the largest building, and found a primitive lunch-counter on the ground floor. Boys and rough-looking men, perched on high stools, shouted out orders to "girls" from eighteen to fifty years old.

"HE'S A COW-BOY."

After waiting a few minutes, Mr. Douglass found a seat, which the boys insisted he should take, and a little later they found two together. The man who left the seat Harry crowded into had on a wide-brimmed felt hat, the edges of which had been perforated all around in openwork.

"He's a cow-boy," Harry whispered in delighted tones.

Meanwhile Philip was trying to attract the attention of the very stout and independent young girl who waited upon that section of the counter. He raised his hand, but she only sneered and remarked, "I see yer!" which brought a roar of laughter from some talkative customers. Soon, however, she condescended to turn an ear in the boys' direction, and they succeeded in ordering two sandwiches and two cups of coffee. When they had finished, Harry said, "Phil, we'll forgive the sandwiches for the sake of the coffee!"

THE GRAND BASIN FROM A BALCONY OF THE ADMINISTRATION BUILDING.

After this hasty supper, Mr. Douglass told them that there were two fine displays that evening—the electric fountains and fireworks on the lake-front.

"Let us see both," said Harry. "There's a place for launches down by the Basin, and the man was yelling out when I came by: 'One launch is going to stay awhile in the Basin, and then going out into the lake,'—I think he said at half-past seven."

Philip looked at his watch. "We're too late by half an hour," he said impatiently.

"Why, no, Philip," said Mr. Douglass. "Our watches show New York time. We have half an hour to spare."

"True," answered the boy. "You are right. I had forgotten that; and, by the way, now is a good time to reset our watches."

So they turned the hands back an hour, and felt thankful that another sixty minutes had been added to the evening.

THE PERISTYLE, EAST END OF THE COURT OF HONOR.

"Now," said Mr. Douglass, "I have a popular motion to present. It is moved that we cease moving, and sit down for a while."

"Seconded and carried!" cried Harry; "and, what's more, I see some chairs"; and he pointed to a row that were strangely vacant, while all around were occupied. The boys walked toward them. Suddenly Harry, who was ahead, came back.

"I don't care to sit down just now," he said; and his companions, coming nearer, saw that the chairs were put over a great break in the pavement to warn people away. They turned to walk toward the boat-landing, and just then the electric fountains in the corners of the Basin nearest the Administration Building began to play. Two foamy domes mounted upward, and were magically tinted in fairy hues, changing and interchanging, rising and retiring, twisting, whirling, and falling in violet, sea-green, pink, purple—it was a tiny convention of tamed rainbows. And, meanwhile, from lofty towers great electric sunbeams fell upon the dome of the Administration Building, and created a cameo against the sky: upon the MacMonnies Fountain, giving it a transfigured snowy loveliness: upon one beautiful group after another, bringing them to vivid life. The beams were at times full of smoke and spray, that gave a shimmering motion to their light.

THE STATUE OF "THE REPUBLIC."

"I have been to a circus," said Harry, "where they had four rings going at once. *That* was bad; but this—this makes me wish I was a spider, with eyes all over me."

"The extra legs would not come in badly, either," said Philip, reflectively.

"Well said!" agreed Mr. Douglass. "Let us get into the little steamer; we can rest there."

They made their way to the landing, bought tickets, stepped aboard just as the boat moved off, and were soon gliding gently out upon the Basin.

VIEW LOOKING NORTH FROM THE DOME OF THE ADMINISTRATION BUILDING—JUST BEFORE SUNSET.

After a short delay to let the passengers view the fountains a little longer, the steamer sped under a bridge, through the great arch of the Peristyle, and made out into the open lake.

To their surprise, the boys found a heavy rolling "sea" on; but as soon as the fireworks began, they forgot all else. Rockets, bombs, showers of fire, floating lights—they came so rapidly that there was a continuous gleam of colored light reflected from the waves. Their launch rounded the fireworks station, and then came to a standstill not far from the Naval exhibit, the model man-of-war "Illinois."

Soon some of the women passengers began to object to the rolling. One Boston woman said: "This is rough; I don't like this at all"; but her bespectacled daughter remarked, as a great bomb of rosy light scattered in a rain of fire, "Well, *I* think it's the smoothest thing I ever saw!" which bit of slang from the prim little Puritan was a great delight to the boys. And as the search-light suddenly sent its beams into a lady's face, she nodded cordially, and said, as if meeting a friend, "How do you do?" Then, turning to her own party, added, "They've just found me."

“THERE WAS ROOM FOR ANOTHER
BOY INSIDE,—AND HARRY
MADE A SKETCH OF IT.”

There were many little incidents that amused Harry exceedingly. One small boy, while boarding the boat, ingeniously contrived to knock his hat overboard; it was at once recovered,—a straw hat has no chance racing a steamboat,—but, like Mr. McGinty, was exceedingly moist. So the pilot went down a dark hatchway and fished out an official cap. The boy put it on. The effect was stunning,—there was room for another boy inside,—and Harry made a sketch of it.

But these trifles were only a relief from the grandeur of the display. Philip said it was the Grandest Grand Transformation Scene imaginable. After a “set piece” had been shown, there was a bombardment of “Fort McHenry,” as they called it—a ship and fort outlined in living fire:

“The rockets’ red glare,

Bombs bursting in air,”

and all the rest of a mimic war. Then, as the fort blew up, the Stars and Stripes flamed forth—“Old Glory”—in lines of light; and, far out upon the lake as they were, the rapturous cheering of the crowds came plainly to their ears.

“Benedict Arnold would never have made that awful break of his if he could have been here to-night,” said Harry, reflectively; then, as Philip began to speak, he said, “Yes, I know he couldn’t have been. Thanks.”

A VIEW FROM THE LION FOUNTAIN.
Looking toward the Grand Basin from a point between Machinery Hall and the Agricultural Building.

Another thing that added wonderfully to the effect of the fireworks was a calliope whistle on some yacht or tug. While the people cheered, the musical director of that steam-tug whistle performed on it with a master hand. It shrieked, it cheered, it yelled, it laughed—whatever song without words could be sung by a steam-whistle was performed with variations. And, queer enough, the effect was exceedingly pleasing. It somehow seemed in accord with the whole spirit of the fête. A bold, generous Western extravagance pervaded the whole affair.

On their way back, they suddenly saw before them a long black hulk. It proved, as they passed it, to be a large yacht lying upon her side, with the masts and yards extending out far over the dark waves.

"How did that happen?" Mr. Douglass asked the pilot, pointing to the wreck.

"It was a collision, sir," replied the pilot; but he gave no particulars.

As the man seemed busy in guiding the swift little steamer, the tutor recalled the old adage about "not talking to the man at the wheel," and asked no further questions.

But the sights of that marvelous American Thousand and One Nights combined were not yet over. As they entered the Basin, their steamer halted to enable them to witness a diving exhibition. On a floating tower stood a man in tights, so lighted up by an electric ray as to be clearly visible from every point around. Raising his hands above his head, he fell thirty-five feet or more into the water. Just as he reached the surface, his hands came swiftly together, and he sank like a plummet. In an instant he was up again, kicking a mass of gleaming spray into the air. Several more "followed their leader."

EVENING ON THE CANAL.

It was a thrilling sight, and, on that cold night, chilled the spectators to the marrow.

As they walked along the edge of the Basin after leaving their launch, the boys greatly admired the statues of animals and men set up near the balustrade. There was a bull, several great bears, a farmer and a draft-horse, a bison (who seemed timid and dwarfed by his surroundings), and others, nearly all modeled with a massive effect that gave them wonderful dignity.

And still the crowd surged to and fro, but now with a decided tendency toward the outlets; the lights flashed and gleamed; the bands played, while the great moon sailed overhead as if it was all a fête to Diana.

Tired as they were when they reached the hotel, the boys could not refrain from talking over some of the principal things they had seen. They did not say much about the buildings, for they knew they should see them again; but they talked of the people, the fireworks, and such queer comments as they had overheard.

"I expected," said Philip, "that we should see a great many foreigners—Turks, Swedes, Germans, all sorts. But I didn't. I saw two or three fellows with fezzes on, but that was about all."

"I noticed that, too," Harry responded. "And I didn't hear much but English spoken. It seems to me that Uncle Sam has done most of this thing himself, and that it's mainly his own boys that are taking it in."

"But it's early days yet," said Philip, with a prodigious yawn, "to make—aw!—comparisons."

"That looks more like late hours than early days," Harry suggested. "Let's turn in."

In a few minutes their clothes were on two chairs, and their heads were sunk into adjacent pillows.

VIEW FROM THE ISLAND AT NIGHT.

CHAPTER III

The Party Separates — Harry Goes to the Battle-Ship — The Government Building — The Convent and the Caravels — The Movable Sidewalk.

SUNDAY proved a welcome relief after the long journey of Saturday, followed by the fête night at the Fair; and they were glad to begin the busy week that was to follow with one restful day apart from bustle and confusion.

At breakfast Monday morning, one of the dishes Mr. Douglass ordered was steak; and, as he sawed through it, he remarked:

"This is tough!"

"But I thought you didn't approve of slang?" Harry inquired, with an air of grave interest.

"I wasn't thinking so much of how I said it as of the fact," Mr. Douglass replied. "But the proverb says that 'shoemakers' children are always the worst shod,' and so we ought to expect poor beef in Chicago, the great beef-market of the continent; but I don't like to waste my strength on mere beef while there is so much before us. What are your plans?"

"If you don't mind," said Harry, after a moment's pause, "I'm going to ask you to let me 'paddle my own canoe.' It is hard for three to keep together in a crowd."

"That's true," Philip agreed; "and especially when one is near-sighted. I think I tried to follow seven different wrong men yesterday."

"Yes," added Harry; "'Follow my leader' is a difficult game to play when we are all leaders and followers at the same time."

"All right," the tutor said. "To-day, then, we will separate. I may not go to the Fair at all, for I have several letters on my mind. Remember, we came away on very short notice. What will you do, Philip?"

"Oh, I think I shall spend a long while in the Art Galleries. It's a good place to go to by one's self, for two people seldom agree about pictures—especially boys."

So, after breakfast, Harry, with a proud feeling of being his own master, set forth by himself. He had a very clear idea of what he wished to do first. He meant to go to the model of a United States man-of-war—the "Illinois." He had read much about the White Squadron, and felt that he would never have so good an opportunity to understand just how a man-of-war was worked.

BUILDING THE BATTLE-SHIP. NOVEMBER, 1891.

He had bought a guide-book to the Fair, and found that the route of the launches would bring him quite near enough to the vessel. But in spite of his singleness of purpose, his thoughts were distracted as soon as he came near the entrance.

He noticed first the clicking of the turnstiles. They revolved so continually, as people passed in, that Harry was reminded of the sound of a watchman's rattle. Next, he caught sight of a white-robed and turbaned Turk standing in line at the "Workmen's Gate," as placidly as if he were in his native Constantinople. Harry's turn to enter at the "Pay Gate" soon came, and he made his way toward the Court of Honor. As he passed the great Liberty Bell, which was chiming musically, he read upon it the words:

A new commandment I give unto you, that ye love one another.

He could not help remembering what followed the ringing of the original Liberty Bell, and he hoped that this, its namesake, would bring peace rather than war—a sober reflection that he recalled later in the day.

To the tune of "Hold the fort, for I am coming," played by a peal of musical bells,—very fittingly, he thought,—Harry began the quick journey that ended when the little launch came to a landing called "The Clambake." When the man called out those words, Harry did not budge; but when the man added, "Here's where yer get off," he rose and abandoned the craft.

On the way there, Harry learned that the ducks in the Lagoon were useful as well as pretty. The pilot said that two or three ducks would do more toward keeping a pond wholesome than six or eight hard-working men.

THE BATTLE-SHIP AS IT LOOKED IN JANUARY, 1892.

He was too early to get upon the "Illinois," and therefore turned back to see the Viking ship. It was not far away; and just in front of it were three armor-plates in which were the imprints left by the great conical shot used in testing them.

Harry had read all about the old Northmen's vessel, and ordinarily could have spent hours in studying her mast, her one crossyard, her awning, the shields along her side—but this was a land of wonders. He looked at the boat only long enough to take a mental snap-shot that he could develop at leisure, and then walked on toward the United States Government Building, passing on his way a company of marines at drill.

But again he was diverted. He turned into the Weather Bureau, and was glad he had done so, because of the wonderful series of photographs he found on the walls. Lightning flashes in streaks and sheets, clouds in storm and wind, moonlight and snow effects, were there, but in impossible numbers. He sighed, wished that he had more leisure, and left. This time he succeeded in getting to the rifled cannon in front of the Government Building, but stopped only long enough to take a sight over one of them.

THE BATTLE-SHIP ON DECORATION DAY, MAY 30, 1893.

He tried to go regularly around the exhibits, but surrendered almost at once. The Patent Office models discouraged him; but the Geological Department!—the great transparent pictures in the windows convinced him that he couldn't (as he once heard a man say) "poss the impossible and scrute the inscrutable."

But he did notice some things.

He sketched the skull of the *Dinoceras mirabile* (and copied the name, too), because he was sure that it was the very ugliest thing in the world. He walked around a section of the big tree from California. He really studied a few life-like and life-size groups showing Indians at work, and wished sincerely that he were Methuselah, and that the Fair would last all his days. It was a petrified Wild West show. He said they were splendid, to a gray-bearded Westerner, who replied emphatically:

"They are *so*—and I have been used to the scoundrels all my life!"

THE UNITED STATES GOVERNMENT BUILDING.

Harry sketched a queer Indian "priest-clown's" head. At first he felt a little afraid to bring out his book and pencil; but he found out that every one had more to do than watch a boy drawing, and before the day was over he drew whatever he chose, entirely forgetting the crowd.

Different things attracted different people. He heard one farmer-looking man say: "My stars, Ma! Look-a here!" and expected to see a marvel. He found only some stuffed chickens. Probably the farmer had never seen fowls stuffed unroasted.

But when he came to the War Department collection he gave up skipping. He *had* to see that. Just at the entrance was a splendid bust of General Sheridan, the face wearing the expression the general must have had when he said at Winchester, "Turn around, boys! We're going back!" Against the windows were more fine transparencies, and the whole floor-space was filled with everything having to do with war and soldiers. Small arms, from a brass blunderbuss to the latest breech-loader—yes, and to the earliest, for there was one Chinese breech-loader of the 14th century.

THE VIKING SHIP.

"Instead of trying to get up new things," said Harry, half aloud, "we ought to go to China and study ancient history."

Harry had a feeling of discouragement in spite of his interest. He had always entertained a vague idea that some day he might give his mind to it and make a big invention—a phonograph or a flying railway, or some little thing like that; but now, when he saw how everything seemed to have been done, and done better than he could have dreamed of—well, he said to himself, "This Fair has spoiled one great inventor, for I would not dare to think there was anything new!"

But then he caught sight of a picture called the "March of Time,"—representing a great procession of soldiers, of generals and veterans,—which restored his good spirits, for right in front, "leading the whole crowd," was a row of rollicking small boys. He was grateful to the artist.

One stand of arms showed muskets—relics of the Civil War—injured by bullets. Into one of them a Confederate bullet had entered to stop a forthcoming shot, and, meeting, they had burst open the barrel. Another had been split into ribbons at the muzzle. There were also relics of the Custer massacre, and a gun recaptured from an Indian after he had tastefully ornamented it with brass-headed nails.

The less bloody side of battle was recalled by General Thomas's "office wagon," the side of which formed a desk when lowered, and revealed some very neat pigeonholes for papers, pens, and red tape. Uniforms and equipment, models of pontoons, artillery, a model of undermining, one by one each claimed the hasty glance that was all any visitor had to spare. A longer look was claimed by an oil painting showing Lieutenant Lockwood's observation of the "Farthest North."

Then Harry returned to the Rotunda, and executed a rapid circular movement, hasty, but full of reverence, toward the cases of Revolutionary and Colonial relics—portraits on ivory, letters, flags, snuff-boxes—an endless array of antiquities. Harry was glad to see one miniature, excellently painted, by Major André; for up to that day he had not thought much of the unfortunate major's drawing, having seen only the well-known "sketch of himself" in pen and ink. Washington's diary was another thing the boy found very interesting: as he said, it was "neat as wax and right as a trivet." Harry wondered whether it wouldn't be fun to keep a diary. This reminded him of the flight of time, and, looking at his watch, he set his face once more toward the "Illinois," for it was after half-past ten.

Many were going that way—and, indeed, in every other. Two small boys who, in sailor suits, strode along the pier like two pygmy admirals, gave him another subject for his sketch-book; but they were but atoms in a long procession, for there was no cessation in the coming and going of visitors all the time he was on the vessel.

TWO LITTLE TARS GOING
TO SEE THE MODEL
OF A MAN-OF-WAR.

He went at once below decks, and came plump up against an ice-machine—"to keep the men cool while in action," he heard a young fellow say. Around the bulkheads were draped flags of all nations, and here and there were hung mess-lockers,—shelves behind wire gratings,—hammocks, neatly varnished kegs for stores, and everything Jack afloat could desire. Upon the lower deck also were glass cases protecting exquisite models of the new cruisers and battle-ships.

"Now, if they'll give me just one of those as my share," said Harry, "I'll go home contented. Anyway, I think I will go to Annapolis and become an officer in the navy."

As if to answer this thought, he came next to the room where the work of the cadets was shown. The splicing, the foot-ball statistics, the fencing foils and masks, were welcomed; but the tables full of text-books and the neat drawings on the walls spoke so plainly of hard study and long hours of work that Harry's determination was somewhat shaken. And, indeed, before he had left the Government Building, a soldier of the regular army, guarding some exhibits, had said to him, "The time for war is over." The man seemed to speak seriously, and then it was that Harry recalled the new Liberty Bell and its inscription. War was not all uniforms and parading.

THE CARAVEL "SANTA MARIA."
The Model of the Flagship of Columbus.

The captain's room and office were most attractive, except that a set of the "Encyclopædia Britannica" seemed out of its element—a British book with a Latin name hardly rhymed with a United States man-of-war.

A courteous officer on the "Illinois" told Harry that people's questions were at times hard to answer. "One man," he said, "looked long at the Howell torpedo, read the labels, and with keen interest wanted to know whether it wasn't a flying machine!"

Harry thought that he might have been told that it was a machine to make other machines fly; but he didn't interrupt the officer, who gave him a clear explanation of a life-buoy hanging in the cabin.

While ascending to the upper deck, he heard a woman say, "Oh, is there another story?" and wished Rudyard Kipling had been there to tell her that it was *quite* another story. But he made his way to the conning-tower, paying heed to the admonition of a mischievous boy who said, "Push, but don't shove."

The conning-tower was hardly big enough to lose one's temper in, but gave the commanding officer full view of his surroundings through tiny slits cut through the solid steel. Electric buttons were convenient to push when he wished the guns, rifles, torpedoes, and other assistants to do the rest.

Leaving the vessel, Harry was again launched back to the other end of the grounds, landing at the Agricultural Building. He passed through this great show-house with his eyes well restrained, but did notice some birds flying about under the lofty roof. He wondered if they had come to study the best methods of securing a living at the farmers' expense, and hoped rather that they wished to know what harmful insects it was best for them to destroy.

After eating lunch at a table in the open air near by, Harry boarded Columbus's "Santa Maria." Coming directly from a modern cruiser, the quaint little cockle-shell was a pathetic witness to the great discoverer's hardships. Harry went into the forecastle, looked at the queer old galley, the swivel-gun, the anchors, and wished that he had been aboard the original on that first westward trip. The modern vessels were scientific, correct, and fine, of course; but somehow Harry would rather have sailed the ocean blue in the days when the galley-fires flared fitfully on these pictured sails.

"GUARDING"
THE "NIÑA."

He skipped the "Pinta" and "Niña," sketching from the shore a sailor on the latter who was "guarding" the little vessel, only reflecting that those on the biggest vessel were better off than their fellows in these two, and went over to the Convent de la Rábida. Harry thought everybody knew about that building; but he met a group of three men, one of whom asked in all earnestness, "That hain't the Fisheries Buildin', is it?" Then the boy remembered how amused the great Napoleon was when they brought to his court a man who had never heard of him, of the Empire, or of the Revolution! Harry wondered whether there might not be in the Fair Grounds a few who hardly recalled having heard of a man named Columbus.

Inside the convent were old charts, pictures, and manuscripts, to which Harry gave but a passing glance. But the open court inside at once gave him a sense of antiquity, and the tropical plants recalled thoughts of distant lands, until he caught sight of a tired man worrying a piece of mince-pie for lunch. He started to go out, and only paused before an old globe whereon the lands were full of odd pictures.

THE NEW "SANTA MARIA" CROSSING THE OCEAN.

"Geography must have been like a book of fairy-stories then," he thought as he left the convent door and came face to face with to-day.

Oh, but he was tired! His legs ached, his back was lame, and he felt like the deacon's "one-hoss shay"—as if he might give out "all at once and nothing first." Seeing in the distance the movable sidewalk, it occurred to him that it was a good place for resting.

The convent had been a little depressing. Others felt the same effect, for he heard one woman say, "I'm glad I'm not a monk"—and then, after a reflective pause—"nor a nun."

As he approached the traveling platform that ran on wheels far out along a pier, this cry met him:

"This way for the movable sidewalk! An all-day ride for five cents—the cheapest thing on the grounds!"

It was irresistible. Harry stepped on the slower platform, then to the quicker one, and dropped into a seat. It proved an excellent change. Out he glided upon the long pier, rested and cooled by the breeze and by the sight of the placid waters, now an opaline green in the afternoon light. Harry thought less of the scene than of his muscles.

"If I wanted to make money at this Fair," he said, "I would put on sale a patent back-rest and double-back-action support; and after the Fair it could be sold to farmers for weeding."

Harry made the round trip, and got off nearly where he started. He did not wish to go back to the hotel, but he could not really enjoy anything more, though so long as he could walk he wanted to see, see, see. Nor was it all seeing; a blind man would have enjoyed that day, so many funny remarks were made, so much music was in the air. Bands played, wheels whirled, people chatted, laughed, and exclaimed.

Everybody seemed happy, perhaps because with all the sight-seeing there went plenty of enjoyable exercise in the clear, bracing September air.

As for Harry, he returned to the hotel healthily weary, but not exhausted.

THE CARAVEL "NIÑA."

“CHOLLY” SPEECHLESS.

CHAPTER IV

Harry Returns to the Hotel — Philip Tells of his Blunder — The Anthropological Building — The Log Cabin—The Alaskan Village — The old Whaling-Ship "Progress" — A Sleepy Audience—Plans.

"A SPLENDID
MEAT SUPPER
FOR 25 CENTS!"

HARRY'S route to his hotel lay through the usual throng of men whose one object in life was to make people buy "a splendid meat supper for twenty-five cents!" His legs felt like stilts, and he walked only because he had become so used to it that he could not stop.

As it was still an hour or two before their usual dinner-time, Harry went up to his room, intending to lie down for a while. When he asked at the counter for the key, the clerk told him that his friend "with the eye-glasses" was already in their room.

Harry found Philip lying on the bed, tired but looking contented.

"Why, you're home early," said Harry, in surprise. "I thought you were going to spend the whole day in the Art Gallery."

"So I was," said Philip, rising to make room for the later arrival. "I started for there. Where have you been?"

"Oh, to the Government Building, the man-of-war, the convent, the caravels—and a lot more," said Harry, as he flung himself upon the bed, first having made himself comfortable by removing his jacket and shoes.

"Did you like it?"

"Like it? Of course I liked it, old slowcoach! But it's too much like being invited to two Thanksgiving dinners—enough is better than two feasts."

"What did you see?" asked Philip.

"See here, Phil," said Harry, smiling mischievously; "do you think I am unable to take a view through a millstone with a hole in it? You needn't think you can put me off by asking questions. What I want to know is why you didn't get to the Art Building. It's not small, you know; you could hardly have passed it without noticing it. Come, out with it, young fellow."

"To tell the truth," said Phil reluctantly, but laughing good-naturedly, "I started out all right, for I looked up the way in the guide-book. I found that the cheapest and quickest plan was to take the railway on the grounds—the Intra—something; yes, the Intramural, which means 'within the wall.'"

A "LOOP" OF THE INTRAMURAL RAILWAY.

"So it does," answered Harry. "Great thing to know Latin. But fire away. I can see there is more in this Fair than a whole brigade of boys can see. Let's hear what you did."

"I took the railway, climbing a lot of steps, and we started. They had signs to tell one where to go, but I couldn't read them very well, and so I went whizzing along without altogether understanding where I was. The stations they called out meant nothing to me, and I had an idea it took a good while to get across the grounds; and—to make it short—I was looking at the view, first one side, toward the hotels, and then the other, toward the Fair

Buildings, and I didn't wake up to my position till the conductor said, 'Going round again, young man?' So I got off, for there I was at the same station I got on at. You see, the conductor had noticed me because I sat near where he stood."

GENERAL VIEW OF THE COURT OF HONOR, LOOKING TOWARD THE LAKE.
Electricity Building. Manufactures and Liberal Arts Building. Peristyle. Agricultural Building. Machinery Hall.

"That's a good one on you!"

"I know it. But I didn't like to start over again, so I came down the steps and walked over across the Court of Honor, along by the Agricultural Building, till I came to the caravels and the convent. I saw those, but so did you. I went next to the Krupp gun exhibit by the lake. That gun was enormous! I believe all the gunners could get inside when it rained. They had a printed label on it, and at first I read it: 'Please set off the gun'; but I knew that wasn't likely, so I went nearer, and found it said 'keep' instead of 'set.' Oh, by the way, just before I went in there, I stopped in the doorway and saw some men diving from a tremendous height, out in the lake,—a much higher tower than the one they dived from on the fête night. I also saw in the Krupp building a pretty little model of the house the great gunmaker lived in when he began."

"What was it like?" asked Harry.

"Oh, just a little square thatched house; but you could see the tiny furniture through the windows. I didn't stay long there, for they were sprinkling the floor, and it was sloppy.

"Next I went into the Leather Exhibit Building; but there were mostly shoes and things there, and I didn't see very much I cared about, except some buckskin suits labeled 'indestructible.' I would have liked one of those, except that it was trimmed with silver lace."

"A little gaudy for you," said Harry.

"Yes, but they were fine. So, seeing signs telling people to go up into the gallery where shoes were being made, I went up. I heard machines making a racket, but all I saw was the backs of the other people who got there first."

"I know," said Harry; "I made a sketch of one of those very exhibits."

"Now, where did I go next? Let me see the map—it's there by you."

"DON'T FAIL TO SEE
THIS EXHIBIT."

Harry passed over the little plan of the grounds, and Philip examined it a moment. Then he went on:

"I see now. I meant to go into the Forestry Building, but on the way I caught sight of some things in the Anthro—"

"—Thropo-pop-o-ological," interrupted Harry. "It's a nice word to say when you're in a hurry."

"Yes," Philip replied, "that was it; so I went in there. And I tell you, you mustn't miss that. It's fine. It has everything in it."

"So they all have," said Harry, hopelessly.

"But there are gymnasium things, and African weapons, all sorts of savage huts and costumes, Greek statues, and views, and bits of work from the prisons and reformatories, showing how boys are drilled and trained to work at trades. But, as usual, I didn't think I could see everything, and so I looked at only a few special cases. One that I remember well showed all sorts of games and puzzles—chess, cards, checkers, halma, pachisi, Indian sticks for throwing like dice, the fifteen puzzle, ring puzzles, wire puzzles, all sorts. The chessmen were splendid. There was one Chinese set there, where the pieces stood on pedestals showing three balls carved one inside the other; and the pieces themselves were little mandarins and things, with faces, and beards, and all. There were enough games in the cases for a boy to learn a new one every day as long as he lived."

"Well?" asked Harry, as Philip paused.

"You don't want me to tell it all, do you?" Philip asked.

"If you will," said Harry. "My ears are the only things about me that are not tired; and I am resting the rest of me."

"All right," said Philip; "I'm willing. I am so full of it, I could talk a week. But I remember now there was one place I went before the Anthropo Building, and that was to a real log cabin, with all the regular old-fashioned things in it; but never mind, I won't go back to that, for I've a lot more to tell, and one thing I know you'll like to hear about specially.

"The next queer thing was the Cliff-dwellers' mound, a big structure, made to look like red rock,—sandstone, maybe,—in which these old Indians in the Southwest used to live. I didn't go into it, although a lot of signs said I ought to; but I saw how the little caves were hollowed out and made into huts, with doors and windows. While I was looking up at it—and it is a high cliff, I tell you!—I saw some nuns all in black climbing over it, and that was a strange sight enough. Out in front were some gray little donkeys,—'burros,' used by the exploring party that found the caves. Then I went on to an old-time distillery, outside of which was a real 'moonshiner's' still that had been captured by the revenue officers.

AN ALASKAN IMAGE.

"Then I came to some Alaskan houses. They were made of great rough slabs, with circular doors cut through the trunks of trees in front. There were little models of them in the Anthropo place, too. In front of them stood those carved totem-poles that we used to see in the physical geography book. I saw by the labels on the models that those poles were meant to tell the history of the man in the house behind each one, and that the more rings there were on the carved man's high hat, the more of a fellow the owner was. There seemed to be lots about whales on them. I suppose capturing a whale was to them like being elected to Congress—maybe harder.

"But, speaking of whales, the next thing I saw was the one I want to tell you specially about. Near the shore there, in what they call the South Pond, was an old-fashioned vessel. I walked over toward it, and read the signs. They said it was a whaling-vessel, a regular old New Bedford whaler. You know about those?"

"I guess I do," said Harry. "I remember reading 'Peter, the Whaler,' and a lot more books like it."

"Well, at first I wasn't going in, for they charged a quarter, and there didn't seem to be many going on board. I was afraid it was not good for anything, but at last I made up my mind to risk twenty-five cents on it. I bought my ticket and climbed the gang-plank. There were just two other men on board besides the sailor in charge."

"'Two other men' is good," remarked Harry.

"You know what I mean. When we got up on deck, the sailor came forward to speak his little piece. He said if we wanted to know how they caught whales he'd tell us. Then he went on with the whole thing, from 'Thar she blows!' down to the cutting up and trying-out of blubber.

"I had often read about it, but I tell you, Harry, it was different to see him hold up the harpoon and the lance, the gun for firing a big harpoon and all. And then we saw the vats for boiling the oil. And he said that out of the whale's head they could dip up whole barrels of clear oil; but the whalebone was the thing they were after nowadays. He said they sometimes got thousands of dollars' worth out of the mouth of one whale.

"After he finished telling about whaling, he invited us below, to see a collection of marine curiosities they had on board. It was a regular old-style ship, with the beams coming close down to your head. All around were cases of curious things—real sailors' oddities: carved teeth and shells, swords from sword-fish, idols, weapons, tools—whatever a sailor could collect. One thing I remember was a harpoon-head that had been bent and twisted around itself by a whale till it looked like a scrawl in a copy-book. Then we went forward to the forecastle, to see the queer little bunks where the men sleep.

THE WHALING-SHIP.

"As I was coming away I bought a little book telling all about the old ship; and it is interesting, I tell you. I haven't read it all yet, but one adventure of that ship the sailor told us about.

"She was out with a big fleet, more than thirty, and she was one of the six that got out from an ice-pack. Then a boat came along after, and reported the rest of the ships as wrecked. The 'Progress'—that's the one I'm telling about—and the other saved vessels threw all their valuable cargo over and took in the poor fellows from the ice. That was what I call square. You can read all about it later. Wouldn't you like to?"

THE WINDMILLS.

There was no answer. Philip turned to look at Harry more closely, and found that the tired boy had fallen fast asleep.

"It's all right for him to go to sleep," said Philip to himself, "but I wish he'd say so when he does it; then I'd know when to stop."

Harry awoke in time for dinner. Mr. Douglass had mailed a number of letters, and he and the boys went to the table together. They found that their walks had given them the best of appetites, and they enjoyed seeing the people at the various tables around them. Mr. Douglass spoke of the excellent appearance made by the crowds, and of their good-humor.

"I was in the Fair Grounds for a short time this afternoon," he said, "and I found myself noticing the people quite as much as the curious things around me. If one ran against another, there was never any ill-humor or crossness. Usually both apologized politely. And yet in many places the crowds were enormous. Again and again I would look ahead of me, and think that I couldn't get through the throng."

THE WOODED ISLAND AT TWILIGHT.

"I noticed that, too," said Philip; "but the spaces are big and the people keep moving, so somehow one always finds a place to pass."

"I tell you what I liked," said Harry; "and that was the little drinking-fountains, where you drop a penny and get a glass of spring water. I found them very welcome."

"And the popcorn!" said Philip. "I don't like it much, but I saw it everywhere. Why, you could smell it in the air sometimes; and every now and then you would hear a crackle-crackle, snap-snap, and there would be a popper full of dancing corn over hot coals."

"Yes, I saw them," said Mr. Douglass. "I found it very interesting to talk to the people. Now and then, when I wished to rest awhile, I would sit down on a bench; and pretty soon a man would come up and drop into a seat beside me. Then, in a minute, one of us would say: 'It's a fine day,' or something of the kind, and, without difficulty, a little talk would begin. One man I met told me he was from Massachusetts, and cultivated tobacco. We had a very pleasant conversation, and gave each other advice about what to see. I think this Fair will do a great deal to bring people together."

"It has already," said Harry, solemnly. "I have seen a number come together even to-day. Where did you go this afternoon, Mr. Douglass?"

"I went to the Art Gallery part of the time," the tutor replied. "But I found it, like the other buildings, too overwhelming—whole rooms full of masterpieces of painting and sculpture; something demanding at least a glance wherever one looked. I found I could not stay long. Walking about and looking upward and downward, and from side to side, is more than any one can endure very long. Besides, the pictures are so good that they make one both think and feel keenly, and that is tiring, too. So after about two hours I surrendered, and came out. I walked along the lake shore during part of my way back, purposely avoiding any sights of especial interest."

"What shall we do to-morrow?" asked Philip.

"Whatever you please," answered the tutor. "Perhaps you might do some photographing, Philip."

"I'd like to, but I hardly know where to begin."

"Suppose," said Harry, "that we all three go to the Midway Plaisance? It's a splendid place to get pictures."

"But I hear," said Philip, "that you can't do very much photographing there. You can get a permit for the Fair Grounds, but the Plaisance exhibits are outside of the Fair's control, and you have to secure special permissions there."

"We might try it," said Mr. Douglass. "You have brought your big kodak, haven't you?"

"Yes, with a new roll of forty-eight films in it," said Philip. "But I shall have to take outdoor scenes, for there's little chance to give time-exposures."

"Well, suppose that we hire chairs to-morrow—the rolling-chairs, you know. One can hire either double chairs or single ones; and then we three will be wheeled out to the Midway Plaisance. There we will let the chairs go, and see what we can do. How do you like it, Harry?"

A LAUNCH-LANDING.

"Oh, it suits me," said Harry. "To tell the truth, I should like to go there soon, for there are so many really foreign scenes in the streets and villages that it may be I can get some good little sketches. At all events, I'd like to go to the Wild Animal show, and see it all. I met a boy to-day, while I was at lunch, who said that it beat any circus he ever saw."

"There are a number of absurd cheap shows on the Midway," said Mr. Douglass, "at least, so the guide-books say; but we can go to the best of them, and let the others alone. I find that the people (as I have told you) are more interesting to me than are most of the exhibits, and the Plaisance is always crowded."

The party had finished dinner, and they went up to their rooms; Philip got out his camera, and looked it over, to be sure all was in working order. Harry laid out his sketch-book and an extra pencil. Mr. Douglass, as he usually did, read over his guide-books, and made up his accounts. But all three went early to bed.

IN FRONT OF THE TRANSPORTATION BUILDING.

IN CAIRO STREET.

CHAPTER V

A Place where Visitors were Scarce — The Rolling-chairs and Guides — Mistaken Kindness — Entering the Plaisance — The Javanese Village — Snap-shots—Cairo Street — The Card-writer — The Soudanese Baby.

THE dauntless three reached the gates next morning at about nine o'clock, and found an even larger crowd than usual. They had to form in line at some distance from the ticket-office, and advanced toward it as slowly as people come out of church. But, as before, good humor was the rule, and, excepting for a few of the weak-minded men who always fight their way through a crowd, there was every effort made to accommodate one another.

Philip heard a woman say, "Why, we are all here to have a good time, and to let other people have the same." It was worst just in passing the wickets, but once through, the trouble was at an end.

"How shall we go toward the Plaisance?" Mr. Douglass asked. He felt that the expedition was undertaken for the boys' pleasure, and wished them to have their own way about it.

"Why don't you take the Intramural, as I did yesterday?" Philip asked. "It will give you and Harry a new view of the grounds, and it's a very short ride to the other end."

"All right," said Harry; "but we must keep our wits about us. I knew a boy once who was carried back to where he started from."

MORNING, OUTSIDE
MAIN ENTRANCE.

For this little dig, Philip gently knocked Harry's hat over his eyes. Harry left the hat untouched until Philip put it back in place. "I don't care how I wear my hat," said Harry, "so long as it is in the very latest style."

As they got on the cars, Mr. Douglass noticed that the gates along the sides were all opened and shut at once by the conductor, and at some stations there were large signs saying, "Don't climb over the gates. They will be opened."

When they were just westward of the Horticultural Building, Harry remarked, "There is no need of getting into the large crowds,—there is plenty of room over there, and only one man has found it worth while to occupy the space."

CHAIR-BOYS AT WORK!

Philip looked where Harry pointed, and saw a workman climbing up a dizzy little stairway half-way to the top of the great glass dome.

"If he should fall through, he'd break a lot of glass," said Philip, reflectively.

They left the railway near the mammoth Building of Manufactures, and walked to its northern entrance. Here Mr. Douglass secured their chairs, the young men who pushed them having the time of starting noted upon cards that they kept neatly inside their caps. Wheeling into line, they rode comfortably along through the parting crowd, Philip carrying his kodak upon his knees, ready for business. He had secured a little card, tied to a string, that permitted him to take pictures "with a four-by-five camera only" for that one day. He had paid two dollars for this privilege, and felt bound to use up his roll of forty-eight exposures.

At first the boys found their chairs a little uncomfortable; but the guides raised the foot-rests until their short legs could reach them, and after that they found the vehicles as comfortable as an arm-chair in a library. It was a bright, clear day—"Just the day for taking snap-shots," Philip said enthusiastically; and everything was plainly outlined by sharp contrasts of light and shade.

As usual, Mr. Douglass began to talk to his guide, and learned that the young man was a college student who was rolling a chair at the Exposition partly for the money he made and partly for the sake of seeing the Fair and the people from all parts of the world. As Mr. Douglass had worked his own way through college, he was able to give his guide some practical advice, which was gratefully received.

"PUCK" BUILDING.

Passing along in front of the Illinois State Building—always conspicuous for its dome—they passed around the Women's Building, and came to the entrance of the mile of curious structures that made up the Midway Plaisance. But before they had come so far, the boys, too, were talking to their guides, who proved to be other college men.

A thing one of them told the boys amused them. The guide said that people, intending to be considerate, would lean far forward when the chair was pushed up a slope. "And that," he said, "brings all their weight on the little guiding-wheels in front, where there are no springs. Then the wheels turn hard, and we have to ask them to sit back. So, you see, the kindest people sometimes give the most trouble."

THE WATER-WHEEL IN THE JAVANESE VILLAGE.

In spite of this warning, when they were ascending the first bridge—one that led across an opening from the Lagoon—both boys leaned forward, as one does in "helping" a horse up hill. But when the guides laughed, the two boys quickly sank back again.

Passing under the elevated railway, they joined the ranks of visitors to the Midway. As they intended to come back another time, they glanced only at the exteriors of most of the buildings, pausing first when they came to the Javanese village. While they rode through the crowd the boys were amused to see the odd glances of those who met them. The luxury of being pushed in a chair was, by many of the newer visitors, considered fitting only for sick people, and their eyes plainly said that two strong, healthy boys should walk. The boys knew this, for they had had the same feeling toward riders during their own first day; the second day's walking, however, entirely changed their views, and they understood that it was a wise economy to save bodily tire when eyes and brain were so busy.

THE JAVANESE MUSICIANS.

"You can ride right into the Javanese village," one of the guides told them; so they bought their tickets and were pushed into the grounds.

Surrounded by a bamboo fence with a lofty gateway was a collection of steep-roofed, grass-thatched, one-story huts. Each had a little veranda in front, and as it was sunny, many of the short, dark-skinned little people sat outdoors at work.

Here Philip expected to get a few more pictures. He had already taken one outside. Leaving the chair in the main roadway, he had gone to the side, where the ground was higher, and had secured a negative (or hoped he had!) showing the crowd thronging the long street between the houses.

But on entering the Javanese village he was told that he could not take pictures without another permit. After a little search and inquiry he found a hut within an inclosure marked "private" and "office." Here he met the superintendent, and was given permission to take views inside the village.

THE JAVANESE BABY.

All the time they were among the Javanese, they had heard a queer musical, liquid pounding. Near the center of the grounds they found the cause. An odd water-wheel of bamboo revolved beneath a stream that flowed from an upright iron pipe, and as this wheel went around it struck short hanging bits of wood that gave forth the musical notes. The wheel had apparently no other purpose than to make a noise—it was a primitive music-box. This was Philip's first camera subject.

His second was also musical. There was a band of musicians playing upon some sweet-sounding metal gongs, and another species of Javanese tom-toms. The musicians smiled encouragingly as Philip waved his camera and gazed through his glasses with eager inquiry, and as soon as they were hard at their music Philip took them.

Another picture he lost. While he was just on the point of pushing the button, a guard clapped one hand over the lens. It was too late to stop, and Philip lost his temper as well as his exposure.

"You can't take pictures here," said the guard.

"The superintendent said I could," said Philip, sharply.

"I beg your pardon," the guard answered politely.

"That's all right," Philip said in a pleasanter tone; "but it doesn't give me back the negative. Next time, please find out before you interfere."

In all the foreign exhibits there were seen many objects with which the boys were only too familiar. For instance, looking through the door of a Javanese hut, Harry saw three cheap American clocks, all in a row; and on the veranda of the same house a man was presiding over a sewing-machine plainly inscribed with a well-known American trade-mark. Nevertheless, the little Javanese themselves were unusual enough: the men wore turbans of figured cotton, a tight-fitting jacket, and then, above their trousers, a short skirt or apron that hung about half-way down the thigh. Some also wore above their turbans wide straw hats.

"THE MAN STOOD UP BESIDE HER, AND THEY WERE PHOTOGRAPHED TOGETHER."

One of the women had a cute little baby in her arms. Philip put a silver coin into the baby's hand, and was allowed to take its picture. But the father held the child. Philip said to Harry, as they walked away, "There's a pretty baby";

then, hearing a gentle chuckle from a motherly-looking woman near him, hastened to add: "For that kind of a baby."

The party had left their chairs in a corner of the village, and were now on foot. As they walked around the inclosure they saw a woman and girl embroidering upon a veranda. The girl was about twelve or thirteen years old, had a tinge of pink in her cheeks, snappy black eyes, and shiny coarse hair.

"HE WAS LAZILY SUNNING HIMSELF."

Philip wanted a picture of her, and, after a talk with the man of the house, at last gained his consent. Philip had a little trouble in making the man comprehend that the girl must come out into the sunshine; but by pointing to the sun and to a side of the hut that was in its full glare, he finally had the little model, blushing prettily, posed in a good situation. The man stood up beside her, and they were photographed together.

No sooner had Philip raised his camera than the sight-seers gathered eagerly about him, until he could hardly find space to reach the button. He pushed it in a hurry, and made his way out. Just a moment after, he secured an even better subject, entirely by accident. Upon another veranda sat a mature Javanese gentleman crouched down upon his heels. He was lazily sunning himself, and Philip leveled the camera and took him before he could say the Javanese for "Jack Robinson." The man opened his blinking eyes at the click of the shutter, but only smiled indulgently, and resumed his basking, like a frog on a log.

A YOUNG LADY FROM JAVA.

Leaving the Javanese village, and ignoring upon their way the appeals of a vender of Java cigarettes—"Ver' sheap! two for five!"—they settled back in their chairs and plunged again into the outside thoroughfare.

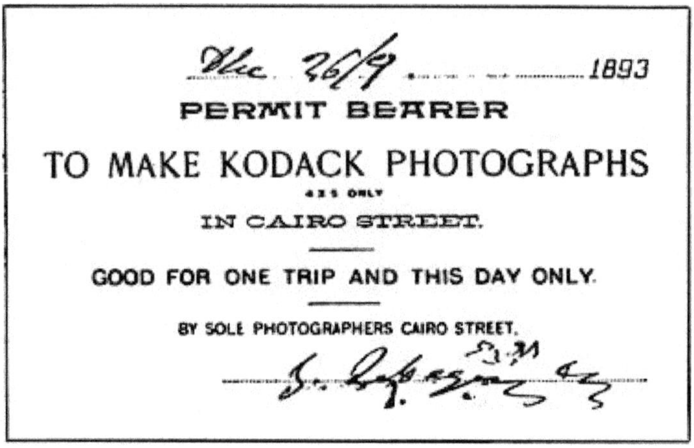

1893

PERMIT BEARER

TO MAKE KODACK PHOTOGRAPHS

IN CAIRO STREET.

GOOD FOR ONE TRIP AND THIS DAY ONLY.

BY SOLE PHOTOGRAPHERS CAIRO STREET.

Mr. Douglass, looking up a little absent-mindedly, saw a sign which he read thus, "Dancing-girl of Damascus now dancing—600 years old." Startled by this marvel, even in that land of enchantment, he turned his head and found that the 600 years referred to the city rather than to the dancer.

"Where would you like to stop now, sir?" asked the guide.

"Suppose we go to Cairo Street, Philip?" said Mr. Douglass. "We can see camels and donkeys and queer buildings without number; and it is said to be a very interesting, genuine exhibit."

They entered the long narrow passage, leaving their chairs outside. Philip's camera was again declared contraband of war, and held in bondage while he "interviewed" the official photographer of the street. He soon returned with the "open sesame" (price $1.00)—another ticket to tie to the camera handle; and they all went forward to view the glories of Cairo.

THE “DONKEY-BOYS.”

It was the liveliest, jolliest place they had yet entered. Donkeys ridden by little boys or little girls came bumping along amid the laughter of the scattering crowd; sneering camels lurched in zigzag courses, carrying giggling girls or grinning men. The camel-riders had the effect of bowing graciously to the crowd, and hung on desperately to the loops of the saddles, as if they were upon bucking broncos. But the most amusing part of camel-riding was the dismounting. The camels went down bows-on at first, and then lowered the hind legs. This process was always sure to bring out little shrieks of dismay from the women, and a burst of laughter from the onlookers.

Philip’s camera was agog with eagerness. He captured a view or two of the picturesque “donkey-boys”—who were stalwart grown men; but when he saw the great nodding camels docilely following their tiny boy-leaders, he made up his mind that the camel was his favorite subject.

He particularly desired to secure a view of the dismounting. Seeing a flight of steps that would enable him to overlook this scene, he put his camera under his arm and wormed his way through the crowd until he had secured an excellent place on an upper step.

From here, by raising the camera high in air, he took a picture over the heads of the spectators, and then rejoined Mr. Douglass and Harry, who were waiting for him across the street near some of the bazaars for the sale of curiosities.

Harry, while waiting, had produced his sketch-book, and made a hasty outline of a street-sweeper who, in turban and baggy trousers, was plying a most prosaic broom and dust-pan.

Just above their heads they read a sign advertising an Arab card-writer, and when Philip returned they began a search for this gentleman, who promised a card in English and Arabic for five cents. It proved to be a difficult matter to find him. Inquiring upon one side of the street, they were directed to the other; and, repeating the question there, were politely sent back again; but soon they caught sight of a ring of people near the middle of the street, gazing down toward the pavement, and there, within, sat the writer.

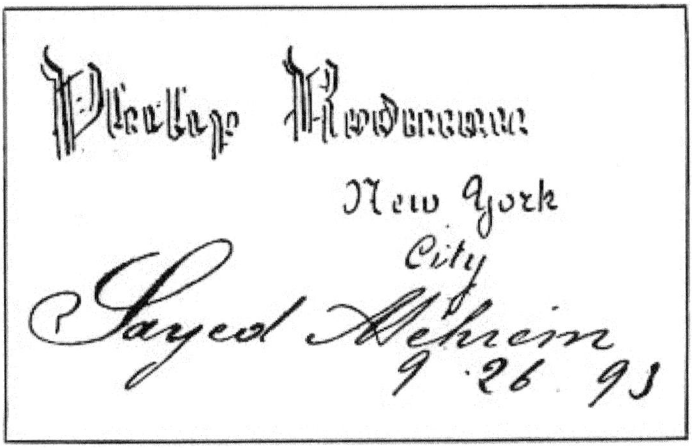

Philip pressed forward with a slip torn from his note-book, on which he had written plainly, "Philip Rodman," putting below, "Please write this name in English and Arabic."

When his turn came, the sharp-featured little writer raised his fezzed head from gazing down upon the inlaid box which served him as a desk, and said:

"You want-a me to write for you—yes?"

"Yes, please," Philip answered.

So the scribe began, like a school-boy reciting his lesson:

"Pheelipe. P, h, i, l, i, and p. Pheelipe. Rodermahn—I write him pretty, in Engleesh, yes; and I vill shade him, yes. R, a capeetal R, o, d, m, a, n. Pheelipe Rodermahn. There. Now, what ceety?"

"Now write it in Arabic, please," said Philip, a little embarrassed by the crowd.

"Pretty soon; in a meenute. You vait. First, what ceety,—vere you leeve?"

IN CAIRO STREET.

"New York," Philip answered.

"All right, all right; I make him ver' preety. N, big N, e, w; Y, a big Y, o, and r, and k. There. Now I write you my own *beautiful* name. See!" and he added his own name with rapid strokes.

"*Très bien!*" said Harry, jokingly.

"Aha, *vous parlez Français*, eh? *Et moi, aussi! Où apprenez-vous le Français?*"

"*À Paris*," said Harry, a little taken aback. "*Je le parle un peu, mais je le comprend.*"

"*Ah, ça va bien! Regardez; voici l'Arabique.*"

Turning the card over, the accomplished scribe traced the graceful curves, and handed Philip the card, saying, "I can write heem as well in four language."

Philip put down two nickels, and waved his hand when the man looked up in surprise.

"*Ah, merci, m'sieu! Je vous remercie, et—au revoir!*"

THE SOUDANESE BABY.

"*Au revoir!*" said Harry; and the three moved away with very kindly feelings toward the clever card-writer.

As they turned toward the further end of the street, an elderly Arab passed them with a stony glare, repeating aloud over and over, "Hello! How-de-do! *Good*-morning! Hello! How-de-do! *Good*-morning!" but paying no attention whatever to any one in particular.

"Now Philip says he'd like to go into the Soudanese Exhibit," said Mr. Douglass, looking at a little plan of the Plaisance. He was a systematic traveler, and always secured a map or plan of each place he visited. They turned into a small inclosure, after buying tickets and seeing them dropped into a battered black tin box (the regular preliminary to all the shows), and found themselves the only visitors in a canvas tent that sheltered a board platform raised a little above the ground. On the platform sat two men and a woman; and about the tent was playing a lively little Soudanese baby—advertised outside as the "Dancing-baby only eighteen months old!"

THE FLOWER-GIRL.

It was to photograph the baby that Philip had come in. But no sooner did the awful black box appear than there was a hubbub.

"No, no!" shrieked the mother, fiercely.

"Nah, nah!" cried the men; and Philip, supposing that he had threatened to interfere with some of their religious scruples, dejectedly lowered his box. But, as they turned away, our innocent travelers quickly had their eyes opened to the true situation.

"One dollar, one dollar!" cried one of the men, following them up. He was tastefully attired in a fez, a long white burnoose (a garment exactly like a nightgown), and red slippers.

Then Harry, who had traveled abroad, felt equal to the situation. He wheeled around with a look of grieved surprise.

"One dollar?" he exclaimed. "Oh, no, no. Twenty-five cents. One quarter."

"No, no. One dollar!" spoke the Soudanese.

"One quarter," insisted the American boy, "or fifty cents for the whole family"; and he waved his arms as if amazed at his own lavish generosity.

"No. Fifty cent for the baby," suggested the dark dickerer.

"Twenty-five in here, fifty if you will take her into the sunshine. Come along," said Harry, starting for the door.

"All-a right!" and the Soudanese made the bargain. For the half-dollar, he conducted the baby to a good light, and let her be taken.

This little tot was as bright as a new cowrie-shell; she had around her waist a dozen rows of tiny dry hoofs taken from some small animal, and these gave her great delight. She crowed and jumped, and rattled at every motion.

"Why, a rattlesnake would be scared to death at such a baby!" said Harry; "and her mother couldn't lose her if she tried. But she couldn't go to church with that thing on—not if she was restless!"

After taking one more picture, the portrait of an Egyptian flower-girl who wandered into the tent, and whose costume, if not her face, was her fortune (at a quarter for every photograph), the explorers waved a final good-by to the rattling baby and turned again into Cairo Street.

Before an attentive circle, just outside the inclosure, an Arab was beginning a performance of trained animals—at least he had a kid poised on a pedestal, and a monkey making ready to ride.

Philip pressed forward to the inner edge of the ring, and leveled the box. He snapped the shutter. Catching the noise, the animal-trainer pulled the kid suddenly down and shook his head with a triumphant grin. Philip moved away, while the bystanders laughed.

"'He laughs best who laughs last,'" thought Philip to himself, as he wound up the exposed film and rejoined his companions.

"'HE LAUGHS BEST WHO LAUGHS LAST.'"

IN CAIRO STREET.

THE FERRIS WHEEL.

CHAPTER VI

The Midway Plaisance Visit continued — Lunch at Old Vienna — The Ferris Wheel — The Ice Railway — The Moorish Palace — The Animal Show.

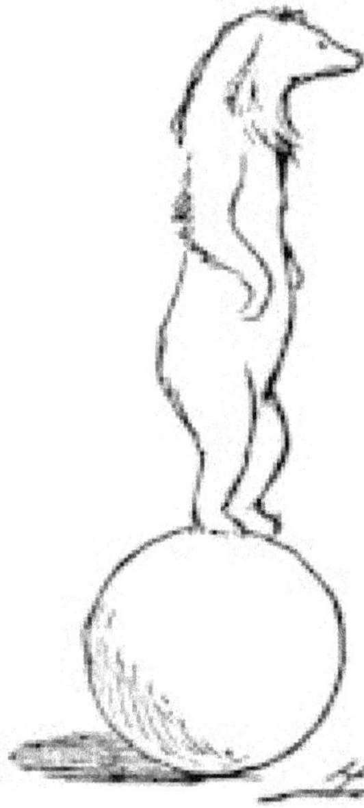

FOR luncheon they turned into "Old Vienna," passing a gorgeous guard in a canary-yellow medieval costume. They found a table under an arbor, and ordered a most unwholesome German lunch. At first Mr. Douglass had trouble in making out the German names of dishes on the bill of fare, and he asked Harry, the traveled member of the party, to read it for him. To his great admiration, the boy translated the items with readiness and accuracy.

"Why, Harry, you are thoroughly up in German eatables, at all events!" he exclaimed.

"It requires only a little careful attention," said Harry, laughing; and, putting down the bill of fare, he showed Mr. Douglass that it had an English translation just opposite the German.

"That is certainly the best system for teaching foreign languages I have seen," Mr. Douglass agreed. "I begin to understand it myself."

After finishing what they could eat,—there was much that they were compelled to abandon,—they sat a few moments over their small cups of coffee, listening to a fine band that played airs from the opera "Carmen."

"When we leave here," said Harry, "suppose we go up in the Ferris Wheel? That gives a splendid view of the whole region, and several people have told me it is one of the best things in the Fair."

"Can I take photographs from it?" asked the camera-bearer.

"We will ask," Mr. Douglass replied.

They were told at the office that they would be permitted to take pictures upon signing a statement that they were not for publication. Philip, however, asked for and obtained a suspension of this condition; and, armed with the permit, they took their place opposite a little door that separated them from the enormous iron spider-web.

In a few minutes the Wheel came slowly to rest, a sliding door was opened, and they entered one of the small cars, of which the Wheel carried some forty suspended within the two great rims. The door was shut, and up they flew, as if in a balloon.

OLD VIENNA.

At first they went completely around without stopping. As they mounted into the air, over two hundred feet, the whole region was mapped out about them. They saw the Fair, the lake, Chicago in the distance,—beneath a veil of hazy smoke,—the Midway, a long white road dotted with its puppet sight-seers. Old Vienna, where they had lunched, dwindled into a toy village.

Philip took several views, but most of them were during the second trip, for then they stopped every now and then to let off and take on passengers, three cars being emptied at a time and at once refilled. He took a view from the Wheel, and a view looking across the Wheel inside.

There was nothing unpleasant in going up or coming down; but when the Wheel stopped, one had the awful thought that something might give way. Now and then came a slight creak or crumble, as if some part was a little strained; but it need not be said that the Wheel did not come down. Neither did any of the cars turn heels over head—that is, floor over roof—as Philip for a moment dreaded. In talking it over afterward, Harry said that his notion was that perhaps the Wheel might stop and leave them up there, and he wondered how they would get down. They came out much gratified with their upward flight, and spoke heartily in praise of the perfect engineering skill shown in the Wheel's construction and operation.

"And do you know, boys," said Mr. Douglass, "the Wheel came here in sections and was put together for the first time on these grounds? It has run smoothly and safely ever since, and is in every way just what its designer meant it to be. He is still a young man, and may some day do even more wonderful things. It is well not to forget that the most difficult engineering feats are not always the ones that seem most wonderful to the public."

GOING INTO THE CARS OF THE FERRIS WHEEL.

"Say," Harry cried out suddenly, pointing southward, "there's something that looks as if it would be good fun."

Philip and Mr. Douglass turned, and saw what looked like an old-fashioned "double-ripper"—a sleigh shooting down and up a long toboggan-slide. As they had no objection to trying it "for the fun of the thing," they went over and bought tickets for ten cents, entitling them to seats in the sled.

FROM THE FERRIS WHEEL—LOOKING EAST.

Once or twice it dashed past them; then it came to a halt, and they all scrambled in, taking their places in the seats, which held three apiece. Then a gong rang, and they were off! Starting slowly, the sledge gradually increased its speed until it met an incline, up which it went more slowly, and would have stopped except that a cable gripped it and hauled it to the top of the hill. Then, again released, the sledge sped down with great rapidity, but was checked by a curve around which it whirled "like all possessed," as a fidgety old lady exclaimed; and indeed the passengers clung tightly to the sides. Around they went again and again, repeating the same experiences until the fourth time, when the car was stopped.

One man, who sat next to Philip, said: "Where you from?"

"New York," Philip answered.

"I from St. Louis!" said the man triumphantly, evidently meaning to call attention to the wonderful fact that the world was small, after all. As they rounded the bend for the third time, the German said:

"I lose my vife!"

"I'm sorry," said Philip, sympathetically.

"Oh, dat's all right," said his talkative companion. "I get her again ven ve stop. She got on other sled. I could not for the crowd. But she vill vait for me; she vill not run avay. She is too good for me, anyhow!"

Philip was relieved that the trouble was not more serious, and after they left the car, the triumphant German pointed to his faithful spouse, saying: "See! I tol' you!"

FROM THE FERRIS WHEEL—LOOKING WEST.

After taking a snap-shot at the moving sled, they left the building, securing at the exit a handful of snow, which was, as the exhibitor claimed, real snow. But he also said it was a souvenir; and as a souvenir it was a failure, unless it was kept in a bottle, for it melted after the manner of all well-conducted snow elsewhere than on high mountain-peaks.

The "Moorish Palace" received their attention next. Upon entering they found themselves in what they considered a very ordinary show. It was a large room having tables and chairs, beer and tobacco-smoke, and a stage where a variety performance took place.

Two young men, in evening dress, were carrying on a dialogue that Harry said was perhaps the most genuine antique in the Plaisance. This dialogue, varied by fair handsprings, lasted longer than the boys cared to stay; so they wandered further into the Moorish mysteries. Groups and figures in wax occupied a large part of the second floor, but the only interesting object the boys saw was a printed sign requesting visitors not to talk to the wax figures. Mr. Douglass's book had informed him that there was a "maze" of mirrors well worth seeing, but in finding this exhibit the party displayed more ingenuity than was shown in the maze itself. "Dime museum" was the boys' well-considered verdict. Turning away, they were attracted by the cry: "Do not fail to see the performance in the great Moorish theater!" Always willing to oblige, the party mustered three dimes, received tickets, and entered at this new door.

"Well, well!" said Mr. Douglass, as he reached the edge of a balcony from which he could look upon the performance. The boys walked forward, supposing that he was expressing surprise. And so he was.

A VIEW THROUGH THE FERRIS WHEEL.

They had paid another admission fee all round for the privilege of entering the gallery of the same room from which they had departed in disgust only a few minutes before. They were grieved rather than angry, and explained their plight to the ticket-seller. He did not let the matter weigh upon his spirits to any extent, nor did he seem much surprised.

"Boys," said Mr. Douglass, as they descended the dusty stairs, "I think that's enough of a maze for me."

When once more in the roadway, they agreed to separate. Mr. Douglass preferred to go back to the Fair; Philip wished to try for a few more photographs, and Harry still kept his faith in the Wild Animal Show.

LOOKING UP AT THE FERRIS WHEEL.

So Philip and Mr. Douglass left him, and Harry walked toward the show.

"Oh, I like the whole business; don't you?" he heard a woman say to a friend; and he was willing to agree so far, if he might except that Moorish maze.

He found a large crowd pressing toward the Animal exhibit, and, buying a ticket at the door, was soon ushered into a very large amphitheater surrounding a circus-ring on a raised platform. Above the ring was a covered cage. Harry made his way toward a number of unoccupied seats, and was surprised that these were so empty while the others were so crowded.

A little boy, coming to collect the tickets, announced: "You can stay here if you like; but you won't see nothing much, for the animals sit around here, and you'll have to look over 'em." So Harry took a better place, near two German gentlemen, one of whom courteously handed him a program, for which there was an extra charge made.

A VIEW TAKEN AT FULL SPEED ON THE ICE RAILWAY.

A scarlet-coated band filled the air with melody, and the show began, introduced by a really blood-curdling roar, such as a healthy and hungry lion gives when he wishes to make an impression. The amphitheater was as full of people as if it had been the only exhibition given that afternoon in Chicago. A baby elephant lumbered in, followed by a large hound and two ponies, and these animals went through a clever performance of marching, wheeling, waltzing, and posing under the direction of a graceful young girl dressed in a close-fitting purple velvet jacket, trousers, and military boots. They were excellently managed, and performed cleverly.

A wild boar came next,—an ugly-faced fellow,—and was put through his feats of hurdle-racing and riding a chariot drawn by another boar. He failed at two hurdles out of three, knocking them over; but was made by the clown, his trainer, to repeat the trick successfully, amid applause. Once the clown made the boar sit down on a high tub, and then cocked a white hat over the animal's ear, giving him a comical appearance.

The succeeding performance was one of the cleverest. A ring-master came in, bringing a small pony whose neck was covered by a thick white pad, and who carried a flat saddle upon his back. Afterward entered a lithe, tawny lioness, who ran cat-like around the ring, and another enormous hound who did little, but was probably an important part of the show.

A SLEEPING LIONESS.

The lioness leaped upon a high platform, and as the pony came around the ring sprang upon his back just as a circus-rider does. Again the lioness leaped from the pony to another platform higher in the air, and awaited the pony's second circuit. It was very exciting to Harry, for the lioness seemed anything but cowed—snarling, raising her whiskers, and showing much spirit.

Harry made up his mind that the hound was brought in as a sort of watch-dog, in case the lioness should show more spirit than the circus-performance demanded; and this idea was strengthened by the presence of these great dogs in nearly every act—but usually as very minor performers.

After the lioness had loped down the sloping passage leading from the ring, attendants came in and removed the carpets and mats used in the circus-performance. They returned with little wooden shelves arranged to hook upon the bars around the great circular cage, and put these in place. Then the lion-tamer entered, not in tights, spangles, armor, or tinsel, but in a dark business-suit that would not have attracted attention in the street.

After him came in a "happy family," as it used to be called in the Barnum days; but not the sort of happy family that would be welcome if it should drop in to spend the evening. First came the dogs, then three bears, two black and one a polar bear, then lions, Bengal tigers, until each of the many little shelves had its occupant.

MEAL-TIME.

These animals were admirably trained, and went through a variety of clever performances. One little black bear—just the sort of little fellow you would expect to see robbing bees of honey, or stealing a squealing little porker from a sty—was led out and invited to show the ladies and gentlemen how well he could walk on a great, blue, rolling ball. As he went forward to begin the act, his lounging gait set all the spectators to laughing, and his whole performance was equally funny, excellently as it was done. When through, he was rewarded by a lump of sugar produced from the ring-master's coat-tail pocket. The same bear also walked the "tight-rope" along a thick bar of wood.

Meanwhile the polar bear acted as a clown. He seemed to find something very interesting about one of the big hounds. During each act, Mr. Polar Bear would leave his place and snuff around Mr. Dog's ears, and paw his neck with the great sharp claws necessary to one who walks much on icebergs and other slippery places. At one time, late in the performance, the bear seemed to conclude that the dog was good to eat, and began to take him in head first. But here the dog's patience gave out, and he howled a gentle protest that sent the polar bear back to his place.

When the little black bear had finished his second act, the ring-master patted him upon the head with a pleasant touch of approval that was kindly and encouraging.

YOUNG LION ASLEEP.

Then the animals changed about: the bear going back to his place, and the Bengal tigers slouching into the ring. A see-saw was put up, and, with a tiger on each end, was rocked to and fro by another black bear—one that had a peculiar white crescent upon his breast. After this the whole company ranged themselves, standing, in a ring, and the big dogs leaped over their backs just as circus-riders leap hurdles.

A chariot came rolling in, a number of the attendants followed, and two tigers were yoked up as if each were "the patient ox obedient to the goad."

The biggest lion, draped in a scarlet cloak and crowned, mounted the chariot, while two hounds rested their fore legs upon the back of the chariot, and around the ring went the gorgeous procession—an animal Emperor making a triumphal procession.

Another pyramid of animals was formed, and then all, set free, went rolling and tumbling about the arena, as their trainer stood among them giving out sugar.

No exhibition of animal-training could have been better, and Harry left the building well satisfied with his afternoon at the Animal show.

Coming out into the Plaisance road once more, Harry started to walk back to the hotel. He had enjoyed the rolling-chair in the morning, but felt freer to go where he chose when he was by himself and on foot. He did not intend to see any more sights than he could help, but the boy had to keep his eyes open to see where he went, and so long as he did not shut his eyes, sights had to be looked at.

In passing the Children's Building, he noticed carefully where it was, as he intended to come back to it soon; then he walked through the "Puck" Building, noticing the color-printing, and the pretty photograph of a child dressed as "Puck," and passed thence across a bridge to the quiet wooded island. His eyes were rested by the soft green tints, and the quiet was very refreshing after the bustle and confusion of the Plaisance. All about were little fairy lamps of different colored glass, arranged in preparation for an illumination of the island that night.

THE POLAR BEAR.

Harry wandered on without attending strictly to his course, and consequently found himself in the middle of the island without any means of crossing to the Manufactures Building. As he wished to walk the length of that building on his way home, he rather reluctantly retraced his weary way to the bridge leading to the Fisheries Building. But this mistake enabled him to warn another party of visitors against the same error, and they followed him over the two bridges to the Manufactures Building.

He was too tired to look at exhibits, and walked doggedly down the long aisle until he came out upon the great Court of Honor. Here he rested a little while, feeling rather dazed, and then walked by the Administration Building in company with many out-going parties quite as weary as he.

A soldier in flaming regimentals passed, carrying a baby in his arms, while the unwarlike wife followed at his side, supporting the officer's heavy sword. This odd exchange of duties was the last thing Harry noticed before he left the gates.

Mr. Douglass came home, and reported that he had spent most of the afternoon in examining the decorations and groups upon the outside of the larger buildings, particularly those upon the Administration Building, as he wished to write some account of it to a friend interested in decorative work.

As to Philip, he resolutely refused to tell the others all about his afternoon except so far as this. He said, "I had some trouble about my camera, and it took me all the afternoon to straighten it out."

Later, his little adventure came out, and shall be told.

"A BUBBLE OF LIGHT."
THE DOME OF THE HORTICULTURAL BUILDING BY NIGHT.

CHAPTER VII

Harry gets a Camera — The State and National Buildings — The Eskimo Village — Snap-shots out of doors — A passing Glance at Horticultural Hall — Doing their Best.

"IN the absence of any special instructions from your father, Harry," said Mr. Douglass, as they walked over toward the entrance of the Fair Grounds on the following morning, "I have so far let you have your own way. I think that Mr. Blake perhaps forgot that his letter of instructions would not arrive at Chicago until we had been here at least a week.

"Now that we have a general idea of the display, of the grounds and their arrangement, I think it would be wise to go at them a little more systematically. What do you think?"

"I should like that better," said Philip. "I feel all the time that we are missing some good things, and seeing poorer ones twice over. Don't you, Harry?"

"I suppose so," Harry answered slowly; "but I find it all too much for me. I find myself thinking more of the people I see than of the show."

"Let us go and see some one part more especially," Mr. Douglass suggested; "some part that we know less about than we have learned of the larger buildings. How would you like to look at some of the larger State buildings?"

"I'd like it," Harry agreed. "But I'll tell you what, while Philip was using his camera yesterday I wanted one 'like sixty.' Why can't I hire one?"

"You can," the tutor answered. "Where do we go to get it, Philip?"

"To the free dark-rooms back of the Horticultural. We can walk there: it isn't far from where we usually go in; or, if you want to go in a new way, we can keep outside until we get to the proper entrance."

THE CENTURY CO'S ROOM IN THE MANUFACTURES AND LIBERAL ARTS BUILDING.

All three were willing; and, keeping outside of the high board fence topped with several lines of barbed wire, they walked on for two or three blocks above the main entrance. The street was lined by booths for the sale of the omnipresent souvenirs—glass paper-weights, watch-charms, canes, lockets, and every sort of cheap knickknack; and these booths were elbowed by temporary shops and stands made to serve for restaurants, fruit-stands, shooting-galleries, tintype-galleries, cake-kitchens,—all the cheap-John establishments that could find room to claim a nickel from the passers-by.

Coming to the entrance they sought, they met a young man in a blue uniform and cap showing that he was an agent of the Photographic Department. Harry paid him two dollars, and received a "hand-camera permit, good for that day only," the date being stamped on it in green ink. They found themselves, after passing the gates, not far from the photographic rooms. Here Harry secured a small, easily handled kodak, upon which Mr. Douglass made a deposit of ten dollars.

"Now," said Harry, "I'll show you how cameras are handled by experts."

"But remember," the tutor reminded him, "that you are here to-day with the intention of going through some State buildings at least. Don't think mainly of taking snap-shots."

"Oh, I won't," Harry replied, more seriously; "I only mean to take pictures of the groups of people here and there—especially the children. Children are always so interesting when they are at a place like this."

Mr. Douglass smiled at the boy's grown-up airs, but said nothing more.

"Come," said Philip, "I want to go over to the Manufactures Building. I saw in my magazine that one could register there, and I'm going to do it. Besides, I haven't been in the galleries of that building yet, and I'd like to go. We won't stay long, and we can meet there if we should separate for a while."

They entered by the north door, climbed into the gallery, and found that some of the periodicals had arranged tasteful little rooms for the accommodation of the public. People entered these small compartments with a homelike feeling that was very pleasant to see. There were tables and chairs, books for the registry of visitors, and glass cases showing magazine- and book-work in full detail, besides many other things connected more or less directly with the subjects of the books and articles published. But, intending to return again, the boys did not linger over the exhibits, pausing only long enough to register their names. Here Mr. Douglass remained to talk to one of the attendants, as he expected some letters to be addressed to him in care of that exhibit, and the boys started together for the National and State buildings.

These filled a large part of the grounds around the great art galleries.

Their first visit was to the house devoted to Great Britain. They marched boldly up to the door, opened it, and stepped inside.

A guard came forward and politely told them that on this morning the building was open only on presentation of a card. The boys turned to go out, but one of the gentlemen in charge—a handsome young Englishman—courteously invited them to go through the rooms. They gladly accepted his invitation and guidance.

"This," he told the young Americans, so politely that for the moment they almost regretted the famous "tea-party" in Boston harbor,—"is called Victoria House by the Queen's own wish. It represents a manor-house of the Tudor period, of about Queen Elizabeth's time; but was made by a Chicago firm." Then he went on to call their attention to the fine ceilings, fireplaces, staircases, and inlaid cabinets; and the boys found the house full of richly carved woodwork and furniture. Of the chairs, one was a model of that in which King Charles sat during his trial in Westminster Hall, and others were quite as well worth attention, among them being chairs designed for the use of Queen Victoria and the Prince of Wales.

VICTORIA HOUSE.

Our sturdy young Americans gazed with becoming reverence upon all this elegance and grandeur, took a few notes of what they had seen, and walked down the steps much gratified by the attention shown them.

"Where next?" said Harry, at the same time taking a deft snap-shot at some little folks in the road before the door.

"Germany comes next," answered Philip, holding up a fluttering map.

"Sprechen Sie deutsch?" said Harry. "If you do, come along."

Entering the imposing German Building, they found at last some of the foreigners as to whom they had been inquiring. No sooner were they inside the door than guttural accents assured them that there were foreigners at the great World's Fair. The hallway was full of German publications, and in a lower story were many religious figures, modeled life-size and colored. Taking a stairway to the right, the boys came to a lofty mechanical clock, called a "Passion Play Clock," because figures, moved by machinery, went through a representation of the crucifixion. They heard a woman say, "Oh, I wish it was going! Don't you?" Then they descended the stairs again, and, returning to the main hall, they noticed a very beautiful stained-glass window at the further end. The middle panel showed Christ walking on the water, and those above and below contained modern steamships. A placard stated that the window was to be presented to the Naval Academy at Annapolis.

Coming out, they were met by a puzzled woman, who inquired in a dazed way:

"Where is that Anthropo—I don't know the name?"

"'Way down at the other end, madam," answered Philip, politely raising his cap. To which the woman responded despairingly, "Oh, my!" and wandered off.

INDIA HOUSE.

"They never get much beyond 'Anthropo,'" said Harry; "and I don't blame them. I heard one of the guides the other day confidently call it 'Anthro-polo-logical' and look proud. But this isn't photography," and he turned his back to the sun and held his camera in readiness. Snap! went the shutter, and then they walked on.

"What did you take?" Philip asked.

"I'm not telling," said Harry, slyly. "I may be new at this business, but at least I know enough to keep dark until the negative is developed. 'Don't count your negatives before they're developed' is my motto as an amateur photographer!"

"Here's the French Building," said Philip; "and isn't it French, though? See the green grass, trees, and fountain in the middle. Let's go in and see it. It is sure to be good."

They found the French Building, as Philip expected, both artistic and interesting. There was an exhibit of transparent photographs on glass, explaining the method of measuring and describing criminals so that they may be always identified after being once in the hands of the police. Here was a panel devoted wholly to queer noses; next came one upon eyes, or chins, or foreheads, each with a line of explanation in French, which Harry translated. Then there was a wax figure before a camera, giving a vivid idea of the way these photographs were secured. A camera upon a very high tripod stood over another figure representing a body found dead—to explain how a picture-record is made of such cases.

THE MASSACHUSETTS STATE BUILDING.

There were specimens of the work of invalids, probably hospital patients, and around another part of the building were large paintings showing views of city squares and streets. The whole building was a proof of the skill of the French in arranging exhibits both sensibly and artistically, so that they would be both easy to view and pleasant to behold.

A room devoted to relics of Lafayette was marked "Closed," for which the boys were sorry. They gave the French-Building a good mark in their notebooks, and went away wishing they could give more time to it—the best proof of excellence.

THE NEW YORK STATE BUILDING.

They had intended next to see the Massachusetts house; but that also was not open, and they went by it on their way to New York's mansion. Entering the great door, they noted first a pavement of tessellated blocks in which were set the signs of the zodiac in brass, finely modeled. Just before them they saw a long line of people crowding toward an enormous book that looked at least half a foot thick. A sign told them that they should register and have their names published in the "Daily Columbian," the Fair paper, as a means of finding old acquaintances.

"Here she goes!" cried Harry, as he took his place at the end of the queue, with Philip next. They could see the book from where they stood, and were much amused, though a little impatient, to see the painstaking efforts of country folks to write a creditable signature. One nice old lady dotted an "i" at least three times, and each time with due deliberation.

As each visitor wrote name, temporary address, and home address, Harry had to wait several minutes for his turn. The result was that he scrawled his own name in a great hurry rather than keep others waiting. Then he went half-way up the stairs, and took two short-time exposures toward the registering crowd. He doubted whether he could get anything worth preserving, but thought he would risk it.

THE OHIO STATE BUILDING.

Then Philip and he went up-stairs to the banqueting-room—a stately apartment of which the boys were patriotically proud. Other rooms—one a colonial drawing-room with an old spinning-wheel, and an old cannon that was "fired at the births and deaths of members of the Rensselaer family," and the other a more modern apartment—fittingly flanked the central apartment.

"Well, we've got a splendid building," remarked Philip, with a sense of satisfaction.

"Yes, sir," said Harry; "the old 'Empire State' always comes up smiling and takes a front seat right next to the band-wagon"; but he, too, was glad that his State was so creditably housed.

Pennsylvania, with a great "Keystone" on the front, was next in their pilgrimage; and here they found the genuine old Philadelphia Liberty Bell occupying the post of honor in the vestibule. Though "marred and bruised by many a thump," the boys gazed upon it with genuine reverence. No American boy could see it without something of the thrill in his veins that is the old bell's due.

THE CALIFORNIA STATE BUILDING.

As they were gazing speechless upon it, a man behind them tried to express what all felt. He began, "That is the bell that—that rang, reverberating down through"—but here words failed him, and he passed silently on, a good though speechless patriot.

Up-stairs they found tired Philadelphians in welcome quiet and seclusion. Even in the "Press-Correspondents' Room" pens moved with Quaker-like dignity over the paper; indeed, one kindly old lady, on looking in at the door, remarked with sympathy; "Ah, yes, I see; people writing home to their friends!"

In another up-stairs room were shown the original charter to William Penn,—a beautiful piece of antique writing,—and the Constitution of the State of Pennsylvania. Attached to the charter was a large wax seal, labeled over two hundred years old.

"Pretty old wax, isn't it?" said a quiet man near Harry.

"Yes—waxing old," the boy replied; but as the man gazed upon him in puzzled surprise, the boy moved off, rather ashamed of his forwardness.

A GROUP OF ESKIMO.[1]

Going out, they noticed General Greene's Revolutionary battle-flag, "baptized in the enemy's smoke" at Bunker Hill. They visited the Ohio Building, also, and then walked toward the Art Gallery; and Harry tried a snap-shot again. This time it was at a chubby youngster who walked before them, carrying two packages of lunch, while his parents walked beside him. Winding up the film, the boys set forward at a rapid pace toward the California Building, pausing only to admire the great logs that formed a foundation to the structure the residents of the new State of Washington had proudly built. The California house was like the pictures of old Spanish Missions; it had an arched doorway, tiled roof, and fine tower.

ESKIMO WOMAN AND CHILDREN.

ESKIMO GROUP WITH SNOW HOUSE.

Though they spent a long time in this building, they were dissatisfied when they came away. There was "so much too much" to see. A relief-map of San Francisco, the knight "Sir Preserved Prunes," the grizzly bear modeled from life, the piece of Laura Keene's skirt showing dark stains where Lincoln's bleeding head had rested, the exhibits of school-work,—drawing, modeled maps, and exercises,—and especially the stage-robbery exhibition made by the Wells Fargo Express Co., delighted both the boys. Then, too, there were paintings—one of Leland Stanford driving the last railroad-spike uniting the Central Pacific and the Union Pacific railroads, several of scenes in midwinter, showing trees in full leaf—in short, the California show sent two Eastern youngsters away full of hearty admiration for the young giant of the far Pacific coast.

THE ESKIMO AND THEIR DOGS.

But by this time, useful knowledge was palling upon the two friends, and they gladly agreed to go back to the Eskimo village, which they had seen just as they turned south toward the Californian mission-house. They deposited two quarters, surrendered two tickets, and walked into Greenland, only to be disappointed in the show. The sledge was upon wheels, which the boys hadn't bargained for—though they hardly expected real ice-floes; and the row of bark huts were dark and commonplace. The natives themselves looked furry and real, and the reindeer and dogs were interesting.

"THE SLEEP OF THE FLOWERS."
A Bas-relief on the Horticultural Building.

Two of the Eskimos, one the well-known young "Prince," held whips in their hands, ready to dislodge coins that might be set up as targets. Harry threw down a five-cent piece. The man stuck it up on edge, and then the whip-cracking resounded through the air. Judging by the number of shots they made unsuccessfully, Harry calculated that a five-dollar bill would have lasted them a month; but he didn't try it.

As they passed the building called "The Bureau of Public Comfort," Harry tried a shot at some people who were eating lunch upon the grass.

Later, he saw a young girl with a kodak making for the middle of one of the bridges, and walked after her, hoping to take a picture of her while she herself was snapping the button. As she leaned against the parapet and leveled the camera, Harry saw that he could get a pretty negative, and himself took the young photographer.

On their way home the boys walked through Horticultural Hall, with its palm-trees, its flowers, and its lofty glass dome. By this time, however, they had learned to see without noticing, and they decided to come back some other day if they had time—a resolution already made in regard to perhaps one hundred and fifty equally absorbing collections.

But there were several fine groups of sculpture. One the boys felt was full of sentiment and beauty; it was called "The Sleep of the Flowers," and meant to be typical of autumn. The drooping of the vegetation and the lethargy of the coming winter were admirably translated into the action of the figures.

"Philip," said Harry, "we ought to see all these groups—everywhere."

"Harry," replied his cousin, "we are doing our level best." And, consoled by this thought, they rejoined Mr. Douglass and went home.

GENERAL VIEW OF THE HORTICULTURAL BUILDING.

AN UNFRAMED PICTURE.

[1] *These photographs of the Eskimo village were made in March, 1893, when there was snow on the ground.*

CHAPTER VIII

What People Said — The Children's Building — The Woman's Building — The Poor Boys' Expensive Lunch — The Life-saving Drill.

A CHAIRLOAD.

"DO you think you are now capable of finding your way around without my help?" Mr. Douglass asked, in planning out the next day's program.

"Yes, sir," Philip answered. "After all, the plan of the Fair is simple enough. It is only after one gets into the buildings that it becomes confusing. Several times I have intended to come out facing one building only to arrive at another. But I can soon set myself right again."

"How about the Intramural?" asked Harry, with affected anxiety. "Have you got that straight yet—or does it still run in a circle?"

"Come, Harry," Mr. Douglass interposed; "Philip has learned better than to go wrong again. What shall you boys do to-day?—I am going to see the Government Building, unless you need me. I should like to see the Patent Office Exhibit."

"I don't know that we shall keep together all day, but Philip and I agreed to see the Children's Building and the Woman's Building, anyway. Besides, there is a life-saving drill on the lake front at half-past two, and perhaps we can get through in time to see that."

Promising to meet again for dinner, the boys left Mr. Douglass to finish breakfast leisurely, and set forth for the upper part of the grounds—the north end.

As they went along, Philip drew out a little note-book and pencil, intending to note down the bits of talk he should overhear from passers-by. He seldom caught more than a scrap, but some of the fragments were queer and suggestive. The first was the expression, "Perfectly magnificent!" Then came a heavy Western man, in a broad felt hat, eagerly telling two friends, "Why, if you was to spend only one second in front of each exhibit—" but they passed on. Then followed these:

THE CHILDREN'S BUILDING.

"Think *I'll* wander around this way?"

"Ain't that it, over there?"

"Get the Orficial Cat-a-logue here."

Entering the Horticultural Building, intending only to walk through it, they heard these:

"You been here, John?"

"Wal, I was just a-lookin' to see."

"Pennsylvania is along here, I guess."

They heard one man assert, "I don't think that it is any good at all!" Whereupon his friend insisted, "Now, you just go along and see."

At a stand where a sharp young woman was selling "ever-pointed" pencils, a man inquired, "What'll I do when the points are all gone?" To which the saleswoman scornfully retorted, "Isn't two years long enough for only ten cents?—but even then you can get new ones at any stationer's."

Coming out of the Horticultural, they caught the words, "The biggest revolver in the world," but never found out whether the speaker was referring to the Ferris Wheel, or to the Equator, or what.

A woman passed by telling her husband about lunching.

"Why, it scares them to death! Twenty-five cents was the cheapest on the bill of fare! But they took it, and they enjoyed it immensely!"

"What do you suppose it was, Harry?" asked Philip, who liked to know all that went on.

"Can't imagine: possibly a watermelon," Harry answered. "It couldn't have been a turkey, judging by the prices we've seen."

Two young girls passed talking about the exhibits. Said one, "I'm not at all sensational over anything." Whereupon the other told her, "Well, I like to get enthused over a thing like this."

THE GYMNASIUM: CHILDREN'S BUILDING.

By the side of the road was a closely cropped and velvety lawn, and over the lawn a patent sprinkler was propelling itself. The water in passing through the pipes set in motion wheels that propelled the little sprinkler slowly over the lawn so as to distribute the water evenly. It was a clever invention, and its utility was evident. Philip and Harry stopped to examine it, but Philip still kept his note-book in hand, and soon had jotted down these entries—speeches made at sight of the little motor:

"Greatest thing I ever saw!"

("Evidently he did not come here by way of Niagara, as Phinney did," remarked Harry.)

"Runs itself—water does it! See?" said one.

"Pretty—good—scheme!" exclaimed another.

"Seen 'em before," came from a third.

"*Ain't* that good?" observed a fourth.

And then Harry and Philip went on; but they talked it over, and concluded that the little sprinkler was a rather independent machine to have set loose on a lawn. Sleeping dogs, and people dozing in hammocks, would have to take their chances.

By this time they had reached the Children's Building, and after admiring the frescoed medallions on the walls, showing children in various foreign costumes, they entered by the main door. First they went up-stairs to the second floor, as they had been invited by the lady in charge to come there at once. Unfortunately, she was not in; but there was plenty in the room to interest them. Upon the walls were large and small drawings, engravings, and photographs of writers known to children or especial favorites of young readers. They saw Miss Louisa M. Alcott ("Jo," of "Little Women"), Hawthorne, Longfellow, Whittier, Mary Mapes Dodge, Thomas Bailey Aldrich ("Tom Bailey," of "The Story of a Bad Boy"), Frank R. Stockton (whose "Jolly Fellowship" was a favorite book of Harry's), Thomas Hughes ("Tom Brown at Oxford" and "Tom Brown at Rugby"), Holmes, Lowell—and ever so many more; but the author of "Billy Butts the Boy Detective" was left out without being missed.

Along the middle of the library ran a glass case showing manuscripts, proof-sheets, and pictures that went to the making of *"The Youth's Companion"* and *"Harper's Young People."* They had already seen a similar display of material for "St. Nicholas" in the publishers' rooms, where they had been the day before. It was a keen pleasure thus to see "how the wheels go round," and to realize that the stories had an existence in pen and ink fresh from the authors' hands.

At one end of the room several bookcases contained books for or about children, from the earliest to the most modern. One book of the seventeenth century was bound in sheepskin and illustrated with odd little woodcuts to show different trades and pursuits. Near these older books were arranged autograph letters from Longfellow, Frank R. Stockton, Palmer Cox, Mrs. Cleveland, Colonel Higginson, Edward Eggleston, Bayard Taylor, George MacDonald, Christina G. Rossetti, Edward Everett Hale, Miss Alcott, Dr. Holmes, Helen Hunt Jackson, D. C. Gilman, and others, of whom Philip and Harry knew more or less. In the library Philip also noticed a picture of Henry D. Thoreau, and reminiscent views of Walden Pond.

THE LIBRARY: CHILDREN'S BUILDING.

Up-stairs, too, was Miss Huntington's "Kitchen-Garden," a school meant to teach the children of poor people in the city how to do well and cheerfully their household work. The little folks sang songs while making beds, setting tables, or sweeping rooms, and learned how to make and how to enjoy a neat home. In another corner was a school where deaf children were reciting as if they could hear, and were reading from the motions of their teacher's lips what she said.

When Philip and Harry went into this room, a big boy was writing upon the blackboard. They heard the teacher tell him to put down five words.

He watched her lips while she spoke, and after some consideration wrote slowly the word "Money." The teacher told him to go on; but, after a long pause, the boy said that he couldn't think of any more. A little girl named Grace put up her hand, showing that she had thought of some; and the boy turned to her, very willing to be helped. So Grace took up his task, and wrote, "Truth, Care, Happy, Mirth"—quite a different kind of words from the sort the boy had chosen. To these short words the pupils added endings, as "Truth-ful=truthful full of truth," "Care-less=careless=without care," defining the words thus made.

TEACHING THE DEAF:
CHILDREN'S BUILDING.

Philip found it hard to remember that these scholars were deaf; but, as the two cousins were leaving the room, they saw at the door a little girl not nearly so far advanced. The teacher was showing her how to pronounce words, touching the child's nose when she did not properly sound the letter "n," and otherwise teaching her the very elements of speech. This sight made it easier for them to understand the difficulties the older pupils had overcome, and they went out with a better idea of the value of sound hearing.

Around the top of one of the rooms was a strange checkered frieze, which, when closely examined, proved to be thousands of card temperance pledges signed by "children of all the world," as the inscription told them. Being red and yellow, the cards made a pretty bit of decoration. Also on the second floor were a kindergarten class-room, with specimens of the work upon the walls; and a class-room for "sloyd," or simple work in wood. But the latter was just then not in use, though there had been classes there not long before.

As they were standing in a corner of the hall, looking at some pictures from children's magazines, drawn by Reginald Birch, Alfred Brennan, and other favorite illustrators, they heard a little boy say:

"Mama, come this way. I want to see the playthings!"

"No; come on. I must see this room," his mother answered; at which the boy whined out:

"Oh-h! you won't let me see a single thing!"

This, if not exaggerated, was certainly a strong statement to be made by a small boy at a World's Fair. To take a child into the Fair and not to let him see a single thing was not only cruel, but even remarkable. Probably the boy overstated it.

Harry and Philip went up on the roof, but found nothing there, and then went down to the ground floor. Here, at one end, was the place where children were deposited while their parents enjoyed the sights at the Fair. One small boy was weeping bitterly, while his father and mother tried to console him. Philip stopped, and the father of the child said, "We were going to leave him here, but he does not seem to like it"; so one boy was not checked.

The boys would have been glad to see these little ones, but the windows and doors were crowded all the time they were in the building; so they gave up the attempt, and only glancing at the Illinois room, spent their last few minutes in watching the children who had come in to exercise.

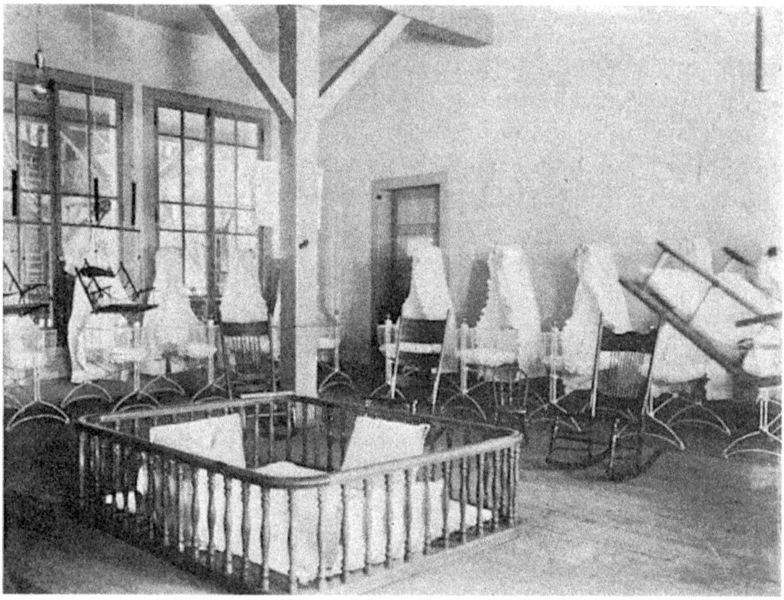

THE NURSERY: CHILDREN'S BUILDING.

The whole central portion down-stairs was fitted up as a gymnasium, and there was a director in attendance to show visiting children how to use the apparatus. There were children jumping, climbing, and swinging, and enjoying themselves keenly. It was open at certain hours every day, and was always filled with young athletes.

THE TOP OF THE WOMAN'S BUILDING.

Feeling that they had now been through the Children's Building, they stepped across to the adjoining exhibit, the Woman's Building, but walked around it half-way, so as to enter at the main entrance. They found the building a larger one than they had expected, and spent more time there than they had thought necessary. Of course there were many things on which no self-respecting boys would waste time—things their sisters might understand, but which they saw nothing in. The embroideries, for instance, were to the boys only pictures; they didn't pretend to say which nation was entitled to the gold medal for needlework. Neither did they pause long before the dressmakers' exhibits. But, still, they found enough in every direction to delay their departure, and it was time for lunch before they were ready to leave. They liked the frescos, particularly that showing the "Lady with the Lamp" among the sick soldiers.

THE WOMAN'S BUILDING.

In the educational exhibit, they heard a little girl exclaim, "Those are mine!" pointing to some drawings; but they did not see much to interest them (in their fastidiousness) except a method of firing colored signal-rockets from guns or pistols; and when they heard a portly woman saying to her friend, "Now, as for me, I would line it with—" they began to rush past everything in the nature of dry goods. An embroidered curtain, showing a combat of dragons, detained Harry long enough for him to declare it "the most mixed up thing he ever saw, for he couldn't untangle t' other dragon from which dragon, and he didn't believe the whole Board of Lady Managers could, either."

A case of dolls showing Dutch, Quaker, and other costumes, the boys were sure girls would like; and while standing beside it, they heard a woman say to her husband: "That doll is dressed the way women dressed when you and I were young." It was a dress such as the boys had seen in pictures of war-times—about 1863.

In one case was some needlework by Queen Victoria, but the ardent inhabitants of our great republic prevented the boys from seeing how deft royalty was with the needle.

"Anyway," said Harry, "she never sat in unwomanly rags plying her needle and thread."

In the art gallery of the Woman's Building the boys noticed only a few of the pictures; "Jean and Jacques," by Marie Bashkirtseff, was one they particularly liked. It showed two little French boys going "unwillingly to school," dressed in their black blouses. Another was a little girl playing hide-and-go-seek behind a low bush. She had a sweet little face and bewitching smile.

They also liked the "Ethnographical Department," where they found all sorts of weapons and utensils from Africa, collected by Mrs. French-Sheldon, the explorer. Harry didn't altogether like the idea of a woman's showing that she explored, just as if she was a Sir Samuel Baker with a great beard, and he consoled himself with the reflection that even Mrs. French-Sheldon probably couldn't whittle a stick.

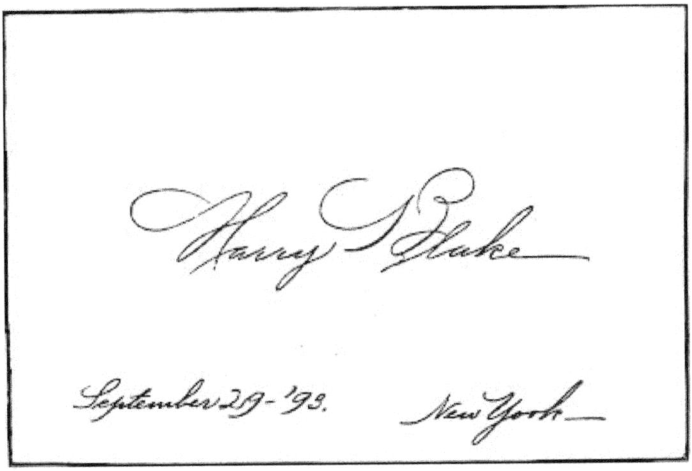

In the gallery were drawings and paintings, among them some by Queen Victoria and other noble amateurs. Harry, owing to the fact that the crowd usually remains below stairs, was able to critically examine the Queen's sketches. The hind legs of one of her dog-drawings particularly delighted him, since they proved beyond question that there is no royal road to animal-drawing. Harry himself had often found the same trouble in drawing the same points, and a warm artistic sympathy welled up in his heart for the great Empress of India in her struggles to conquer animal-drawing. When, in the same gallery, he saw some drawings by Mary Hallock Foote, an artist whose works he admired, he believed that he would rather be a plain American who could draw than a crowned queen who did very well considering how busy she was with state matters.

They glanced into the stately California room, upon the floor of which was a great grizzly-bear rug, and then made up their minds that it was time to be lunching if they intended to see the life-saving crew at work. But on their way out, they stopped long enough for Harry to have his name written by a woman card-writer, who used a pen set "skew-shaw" on its handle. She added his residence—the State only—and the date. It cost him five cents, but he felt that Philip was no longer one ahead of him.

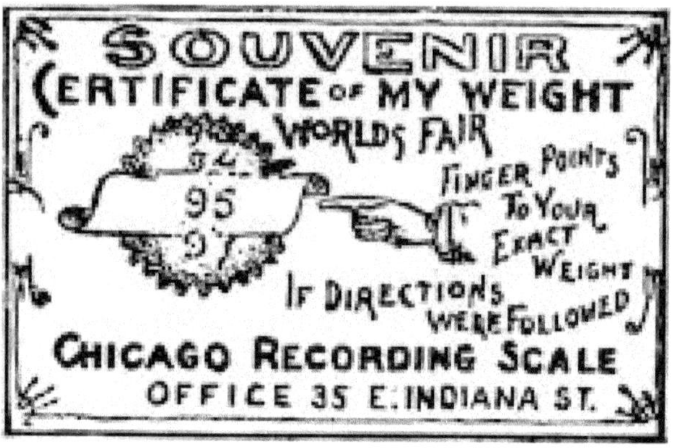

Philip saw a machine marked "Music, Fortune, Weight," with the usual request about dropping a nickel. He stood on the platform, and dropped the nickel. The machine played "The Sweet By and By," and shoved out a ticket upon one side of which was stamped his weight, "95," and upon the other was, "You will soon receive a fortune from across the sea."

They walked between the State buildings over toward the lake, intending to take lunch somewhere nearer the shore. When in front of Ohio's Building, with its projecting portico, they stopped to look at the great statue in front. A woman's figure upon a lofty pedestal raises her arms proudly as if to call the attention of all the world. Around the pedestal, like a row of bad boys sent to stand against the wall for whispering, are a ring of Ohio's great men, including Grant, Garfield, and Stanton. In prominent letters around the pedestal are the words, "These are my jewels." While the boys were looking at this little piece of justifiable brag, two women came along, and paused beside them.

"'These—are—my,'" then moving a little further,—"'jewels.' Hum! Yes; of course. Those are the words that Queen Isabella said to Columbus, you know, when she gave him her jewels to fit out his ships." Both then walked away, enriched with the spoils of history.

Philip and Harry looked at each other, but made no remarks. Their minds were busy in replacing the State of Ohio, Queen Isabella, and the noble Cornelia in the niches from which they had been so rudely torn. In some ways, that was the most remarkable exhibit they met that day at the Fair.

At the same table where they had lunch, a young fellow sat down with two little boys. They looked poor.

"What will you have, Johnny?" the eldest asked one of the little fellows.

"Bread 'n' butter."

"That's cheap," the eldest said; and, after a little more talk, they ordered fried sweet-potatoes.

"Nothing else?" the waiter said.

"Nothing else."

When they were through, the waiter was asked "How much?"

"Ninety cents."

Then there was silence, while the big boy fumbled in his pockets. Four tiny bits of sweet-potato, bread and butter—ninety cents. It was hard, and Harry spoke to the waiter about it.

"I can't help it," said the waiter. "It's the rule." So the bill was paid. It was another interesting exhibit.

They gladly left this restaurant, and made their way out into the honest breeze from the lake, taking their places upon the shore so as to see the life-saving drill. It proved well worth coming to see.

The first sign of life was two rows of white-jacketed men that filed out through the dense crowd which lined the lake shore. The lake was rough and spray shot high into the air as the waves rolled against the breakwater. But the men rushed the boat down the beach, and steered by one who stood in the sternsheets holding a long oar astern of the boat they made their way out to a mast that rose from above the water's surface to represent the mast of a wrecked vessel. It was a struggle, but they finally reached the mast, and one man and a boy got out of the boat and stood upon a small platform not far above the waves.

With even more difficulty the boat returned to the shore; and, after some delay, probably to arrange the life-line and mortar, "bang" went the shot, and the line was carried by the missile fairly across a boom projecting from the mast. Then the man at the mast hauled in this light line until it brought him a heavier one; and again he hauled until he had the end of a cable that came from the crew on shore. This he rove through a block upon the mast, and made it fast. It was made taut by the crew of life-savers, and out along this thread of salvation rolled the "breeches buoy," looking like a Quaker's hat turned upside down.

Into the breeches the boy put his legs, and was hauled ashore by a light line.

Just as the boy came near shore, his legs came so near the water that he drew them up, frog-like, and the great crowd of spectators laughed and cheered. Again the little buoy and breeches traveled out to the mast. But the man out there had noticed the boy's gyrations, and seated himself on top of the buoy.

"You bet your neck he ain't goin' to run any chances of getting *his* legs wet!" cried a very appreciative young man; and the wisdom of the remark far exceeded its elegance.

After the man was landed, the buoy traveled again to the mast and struck against the block there. Automatically, the cable was released and hauled ashore, and the same bolt released the rope, dropped a sign that had been rolled up like a map, and every one could read in plain black letters the words: "DRILL FINISHED."

Before the boys started for the "exit," it began to rain, and immediately there was a fine exhibit of umbrellas from every State in the Union. To keep dry the boys walked the whole length of the Manufactures Building.

Harry timed their walk, and counted his steps. He was going slowly, with no desire to break or make a record. It took about 720 steps to go the full length of the largest building in the Fair, and the walk lasted nine minutes.

Before they went to bed that night Harry was told of a remark overheard by one of his friends. It was made by a tired old lady, who had come out of a large building and arrived unexpectedly in a strange and distant quarter of the grounds:

"Well!" she exclaimed, "when they planned this Fair, they put these buildings so that, wherever you come out, you ain't anywhere nearer any thing in particular!"

THE LIFE-SAVING BOAT.

JUST FROM THE RANCH.

CHAPTER IX

The Manufactures and Liberal Arts Building — A Rainy Day — A Systematic Start — "Irish Day" — Harry Strikes — Some Minor Exhibits — The Few Things They Saw — The Elevator to the Roof.

A DISTORTING MIRROR.

IN the old days the navigators at first crept from headland to headland; then from island to island, and at last Christopher Colon, the intrepid hero of the fifteenth century, conceived the idea of sailing boldly forth into the unknown, secure in his faith in himself and in his fortunes. At least so Philip said in one of his old school compositions. And the boys, having at first touched here and there the points of interest, then took up a few of the outlying State buildings; but now they intended, as Harry boldly put it, to see the elephant from beak to tail-feathers. That is, they planned to enter the Building for Manufactures and the Liberal Arts.

"This mammoth structure," began Harry at breakfast, in the tone of a dime-museum lecturer, "is steen hundred feet long, and even wider; and is provided with wings on all four corners, if not oftener. It contains the complete contents of the building, and various souvenirs and nickel-in-the-slot machines which are not reckoned in the table of contents. Little boys have been seen to enter at one end, and old men to come out at the other, besides those who went up on the roof; so you can draw your own conclusions."

"That's all very well, Harry," said Mr. Douglass; "but have you been up in that elevator?"

"No, sir."

"When you go, see if you feel like joking," Mr. Douglass went on. "I went up in it to-day."

"It's raining hard," said Philip, looking out of the window, "and I haven't any rubbers."

GENERAL VIEW OF BUILDING FOR MANUFACTURES AND THE LIBERAL ARTS.

"Nor I," said Harry; "but I didn't pack the satchels."

"Well, we forgot them," said Mr. Douglass; "so I suppose Philip and I ought to go out and buy some for the whole party."

"Let's all go together," Harry suggested.

They set forth, keeping a good lookout in all directions for anything like a shoe-store. So near the Fair it would have been easier to find the rarest thing in the world than simply a pair of rubber shoes. But finally they came to a shoe-store, and discovered that they were not the only little boys who had failed to imitate the little Peterkins in providing themselves with rubber boots. There was a long line of customers extending out upon what was called the sidewalk, good-naturedly awaiting their turns to be shod. They took their places at the end, and when rather moist, were admitted to the store in a chosen batch of six. They had to wait on themselves, and picked up the first thing that came. Mr. Douglass's first catch proved to be infants' overshoes, but Philip found a pair that Mr. Douglass could wear. The proprietor told them to help themselves, and make themselves at home.

PORCH OF MANUFACTURES AND LIBERAL ARTS BUILDING.

"All right," said one of the customers; "we've all gotten acquainted while waiting on one another out here."

Once well insulated from the ground, they turned the rattling stile at the entrance to the Fair, and picked their way over the mud that was like gray paint and nearly as sticky. The program declared that it was "Irish Day," and the same fact shone out from many a noble breast, and many a proud coat-lapel; for green badges flourished like bay-trees in the spring, and the shamrock bloomed despite the stormy skies.

As they crossed a bridge from the Electricity Building to that of the Manufactures, they noticed that the dome upon the Illinois State Building was as unsubstantial and shadowy as a ghost. The crowd talked much less than usual, and there was little laughter. A number of French sailors passed them, but even their busy tongues were for once silent. The boys were glad to get into the great building, for it was here and there lighted by electric lamps, and the gaily colored exhibits diverted their minds from the gray and cloudy sky.

ANOTHER VIEW OF THE MANUFACTURES AND LIBERAL ARTS BUILDING.

"Boys," said Mr. Douglass, as they paused in front of the two elephant-tusks that rested before the Siam exhibit, "if you prefer it, I will go with you; but, to be frank, I am inclined to think you would rather go by yourselves. Whatever you look at, you are sure to learn something, even unconsciously. And I am not ashamed to say that no one man can explain even to boys of your age a thousandth part of what we see here."

"Mr. Douglass," said Philip, "I really think we'd rather go alone, if you don't mind." "Very well," the tutor replied; "I shall probably stay in this building, too, but it is not likely we shall meet. At about half-past one come over to the bridge that leads to the lower end of the wooded island, and I'll take you to lunch. *Au revoir*!"

"*Auf Wiedersehen!*" Harry replied; and turning to Philip he said warmly, "Mr. Douglass is a good fellow,—there's no 'Uncle George and Rollo' about him."

"I think he's right, too," said Philip. "If he was to try to tell us about things here, we couldn't listen if we wanted to: there's too much to see."

"Well, he's having a good time, too," said Harry. "It's a good idea to take your tutor to Chicago and improve his mind. Where shall we begin?"

"We'll go over into the publishers' corner," said Philip, pointing to the little map in his guide-book; "and we'll take the galleries first."

FROM A WINDOW IN THE MANUFACTURES AND LIBERAL ARTS BUILDING LOOKING NORTHWEST.

They walked toward that end of the building, but could not help seeing some things on their way. One was a group of curved mirrors that gave distorted and very laughable reflections. Another was a fine display of daggers, pins, and other jewelry, inlaid in gold. Harry took a fancy to one bonnet-pin (he thought it was), the top of which was a dainty sword-hilt. He priced it, and left it there: it was twelve dollars. The boys saw a placard upon one tiny dagger saying it was sold to Miss Blank, and they wondered whether she bought it for a paper-knife, or intended to become a vivandière.

THE ARTS OF WAR.
A MURAL PAINTING IN ONE OF THE PORCHES OF THE MANUFACTURES AND LIBERAL ARTS BUILDING.

ONE OF THE DOMES OF THE MANUFACTURES AND LIBERAL ARTS BUILDING. PAINTED BY J. CARROLL BECKWITH.

On reaching the gallery, they first went through the publishers' exhibits, finding original manuscripts and drawings, collections of finely bound books, and courteous treatment everywhere.

"I think," said Philip, as they came out of the last of these rooms, "that the publishers are all very polite to the public."

"Ah," Harry replied, with a wise shake of the head, "they have to be. If they weren't, why we'd just turn around and say, 'Here, you, stop my subscription!' and then where would they be? You see, a man can't get along without food, and clothes, and things like that, but he needn't read if he doesn't want to—he can just spend his time over advertisements, and signs, and things people give away."

"Wouldn't that be nice to have in schools?" said Philip, pointing to a big map of the United States upon the wall, nearly twenty-five feet high.

PART OF GROUP ABOVE MAIN ENTRANCE OF MANUFACTURES AND LIBERAL ARTS BUILDING.

"Very," said Harry; "they'd have to spread it out in the yard, and then the teacher would say, 'Johnny, run out and find Oshkosh, and don't run too fast or you'll tire yourself before you get there!'"

On the opposite wall was "the largest photograph in the world," a very long but uninteresting picture of those words with figures of real people leaning on the letters.

An old man came by, saw the sign, wondered "*where* that photograph was," and walked all around the gallery trying to find out. It was hardly a successful exhibit, but it was only to attract attention—there was a good display of regular work near it.

The boys at first stopped everywhere; but soon they began to remember what a task was before them, and they quickened their pace.

Philip entered but few items in his note-book, and among them was a booth entirely covered outside with ordinary playing-cards, which gave it an Eastern effect. One object that called for more than a glance was an old English clock—the Earl of Pembroke's clock; it was set in a high case of carved wood, most elaborate in design and executed with minute skill and care. They saw also a show-case that imitated a great trunk some fifteen feet high, with glass sides. But they were making slow progress, and hurried on until they reached a carved altar made by the inmates of St. Joseph's Orphan Home—a piece of woodwork worthy of any hand.

Then began a long array of exhibits meant to illustrate the progress of scholars in lessons and manual arts. Each compartment was allotted to a certain school. For a few rooms the boys kept seriously at work examining drawings, carvings, forgings, and compositions; but soon they heard a rollicking pianist down-stairs dashing off "St. Patrick's Day in the Morning," and it brought memories of home to their minds. A lively jig-step was heard, followed by clapping and cheering.

"See here, Phil," Harry broke out, facing about, "it may not be St. Patrick's Day, but it certainly is Saturday, and I'm not going to be hood-winked into school work to-day. If there are any more compositions, kindergartens, and maps drawn by Bertie Wilhelmina Marie Jones, you may see them if you like. I am going to skip them."

"I've seen enough; we'll never get through this way," said Philip, looking despairingly at his watch. "So we'll go on to something else."

"Good-morning, boys," said a slightly husky voice.

"Good-morning, sir," they replied, turning to find an old Irishman, a respectable quiet-looking man.

"I tell you this is a very wonderful show," he went on, evidently feeling that he must talk to some one. And from that beginning he went on to tell them that he was over sixty years old, had come to America in 1847, and had gone West by the Erie Canal, soon after.

"Boys," said he, impressively, "you've no idea of what a country you live in. I've lived to see wonders in the last thirty years, and they've changed the whole world, so they have. You can have no idea of it, not as I have. And it's not in the East or in Chicago alone: it's in the whole land. And there'll be no telling what a country it'll be. I'm over sixty, and I went out forty years ago and took up a hundred and sixty acres of bare land, and now there's people all around me: Norwegians, many of them; and it's good people and good neighbors they are!"

The boys were impressed by the seriousness of the old man's talk.

"You are Irish?" asked Philip.

"Of course," he said, with a smile; and throwing open his overcoat he displayed a badge big enough to prove anything. They parted with mutual wishes of "good luck."

Since Harry had refused to go further into the exhibits of school work, they went down to the main floor, and walked from the southwest corner northward. As in the other buildings they had visited, they found along the walls little stands where young women had on sale penholders, souvenir coins, shell-boxes, necklaces—cheap trinkets of all sorts. For the first few days the boys had gone to see what was shown at these booths; but soon they found there was pretty much the same stock everywhere, and walked by indifferently. They had bought, however, a few things—one a little shield showing the arms Queen Isabella granted to Columbus.

"—AND THE CAT CAME BACK."

Against the wall about half-way up toward the north end were several "graphophones"—contrivances something like Edison's phonograph. On dropping a nickel and hooking two hard-rubber tubes into the ears, one might hear instrumental music or songs. A small boy tried one of those machines while Harry and Philip looked on. The tubes were adjusted, and he stood gravely awaiting the result. A smile began to dawn on his lips. It spread widely. His mouth opened; he giggled aloud; he kept on giggling with his eyes closing through pure joy.

Harry tried the machine and found that it was repeating a comic singer's rendering of "The Cat Came Back," and he grinned quite as widely as the small boy had done, and afterward sketched the scene with full sympathy.

"That's a great invention for invalids," said Philip, thoughtfully.

"Yes," said Harry, warmly; "think how it would soothe a restless invalid during a long night to hear one of those machines grind out 'The Cat Came Back!'"

A JAPANESE
CARVING.

"Well, it would," said Philip, as soberly as he could. "You couldn't be sad while listening to that song."

Just as they were leaving, they saw a mother and child listening to the same graphophone, each having one ear to an end of the branched tube. "I don't know," said Philip, "whether that's quite honest."

The exhibit of a well-known manufacturer of steel pens had in the center of it a pen fully six feet long, apparently quite as huge an affair in its own way as the building. The boys stopped at this, but perhaps at another time they would have passed that by and looked at things they now ignored. There was so much it made them particular. If a display was not brightly lighted, or was at all crowded, or required a few extra steps, it was left unvisited. Knowing they could see only a few things, they simply walked along, and let the exhibits show themselves.

THE HUNTERS' CAMP.

There passed them in rolling chairs an old minister and his wife, and Harry made up his story about them. He imagined one of the deacons going to consult with the elders, saying, "The Parson wants to go to the World's Fair. He hasn't said so exactly; but I can see he does. He reads all about it, and he talks about it—tells how big the buildings are, and all that. Can't we send him?" There may have been no truth in all this, but it gave Harry great pleasure to see the old couple's enjoyment. Coming to the upper, or north, end of the building, they found the exhibits of stonework, ironwork, paints, varnishes, and so on. But they turned back to see the exquisite work of artistic Japan. Here were ivories, pottery, metalwork, embroidery, odd carving (one little bear, a grotesque figure, Harry stopped to sketch)—all designed and executed in perfection. The boys spent a long time here, and left dissatisfied. It was time to meet Mr. Douglass on the bridge, and they raised umbrellas, tramped through the mud, and, finding the tutor waiting for them, were soon on the way to the Horticultural Building, where they lunched at a restaurant on the second floor.

INTERIOR OF THE MANUFACTURES AND LIBERAL ARTS BUILDING—SHOWING THE ELEVATORS.

"Where did you go, Mr. Douglass?" Philip asked.

"I went to the other end of the grounds, to the Anthropological Building. I heard there was a set of apparatus for measuring nerve-force, mental-power, and so on, which would be applied to a visitor. I went through the process, and found it very interesting, though it took a long time."

"Did you notice the Hunters' Camp and the Australian Bark Hut near the bridge we came over?" Harry asked.

THE FIRE-BOAT "FIRE QUEEN."

"Yes; and went into both," said Mr. Douglass. "How well they contrast with these enormous, complex show-buildings, reminding us how much that is shown here is not necessary to life or happiness! After lunch I'll go back with you to the main building, and we'll ride up to the roof."

Walking back, they noticed on the railings of the bridge a life-preserver and line, hung ready to be thrown at once to any one who might fall in. They also saw the "Fire Queen," a steamboat fire-engine, lying ready for service by the same bridge. "That shows," Mr. Douglass said, "how carefully everything here has been thought out."

Returning to the big building, they went through the silversmiths' and jewelers' exhibits, which were rich and elegant without being gaudy or tiresome. There were great crowds here—and they saw only a few of the pieces of silverware and jewelry. The Tiffany Glass Company's beautiful chapel they pronounced one of the successes of the Fair, and just opposite they stopped to examine many watches, watch-movements, and the machines that made them.

Coming to the elevators, they bought tickets and entered, without particular thought about the trip. The door was closed, and the elevator began its upward journey. Until it was near the top Harry didn't look down. All at once he turned his head and saw the awful depth, where tiny figures moved noiselessly about. He was not an over-sensitive boy, but for the minute the sensation was one of appalled horror. It was not fear—he had no dread that the elevator would fall; he only felt the terrible height. It was an instinctive human shrinking before the immensity of space.

He turned away, and did not recover for several minutes. He had no inclination to joke, and, indeed, for a while he could hardly summon courage to step upon the board walk that led out upon the roof. Perhaps Mr. Douglass and Philip had somewhat similar feelings, for all three sat down upon a bench outside, and did not attempt to walk around the roof.

That one moment of dread did more to make the boys understand what a monster building they were in, than columns of figures, comparisons, and statistics could have accomplished. About smaller buildings one can reason; but this can be comprehended only when one is awed by its immensity.

THE ROOF WALK, MANUFACTURES AND LIBERAL ARTS BUILDING.

CHAPTER X

Philip at the Art Galleries — The usual Discouragement — Walking Home — The "Santa Maria" under Sail.

IN THE ART
GALLERY.

THE next morning Philip decided that whatever the others were going to do, he had left the Art Galleries alone long enough, and that he would spend a part of his day wandering among pictures and statues.

He walked from the southern end, where he had entered, along the whole length of the grounds. When he came to the bridge crossing the waterway between the North Pond and the Lagoon, he met two ladies evidently at a loss whether to turn to the right or the left in order to reach the Art Galleries. Raising his hat, he drew forth the map and showed them that they could go as well one way as the other; and then he walked on, himself turning to the right. As he went along a path that led him around some of the smaller National Buildings, he saw a little grove of trees surrounding a boulder built of staff. Along the top of this great rock was a figure, also of staff, representing a lioness with the head and shoulders of a woman—a sort of sphinx; a Cupid was whispering to her, and she had an expression half of amusement and half of malice. There was no legend or inscription attached to the piece of sculpture, and all were left to make their own interpretation of the allegory.

Considering the wealth of art stored up in the winged temple to which he was going, Philip did not dare to waste any time in reaching his goal; but first he drew out his little guide-book, and examining the plan that showed where the pictures of each nation were grouped, he decided to begin with the French section—that is, with the east wing.

AN ARTIST'S VIEW OF THE FINE ARTS BUILDING.

He mounted the great steps, flanked by lions, and found himself at once surrounded by pictures on all four walls of a square room whose curtained doorways led to similar treasures beyond. Like all the world when in a picture-gallery, he did not see how he could examine the collection systematically. He was too much interested. Perhaps he would make up his mind to begin at the right-hand corner, and would march resolutely in that direction. Upon the way he would catch sight of a thrilling battle-scene or a lovely face, and would pause, become fascinated, and lose all recollection of his plan of campaign.

AN INTERIOR VIEW
OF THE DOME OF THE
FINE ARTS BUILDING.

After standing bewildered for a minute or two, Philip turned to look especially at a large painting showing Christ talking to the woman at the well, a beautiful and dignified piece of work, emphasizing the serenity and solemnity of the scene. Philip felt that this picture had put him in a receptive frame of mind, such as one should have when listening to a sermon; and not long afterward came a series of four well-known pictures, "The Prodigal Son," by Tissot, to preach the sermon. They represented a modern reading of the parable, showing the father bidding his son farewell; the son in anything but good company while absent; the return—a touching picture, showing the old father leaning to raise the young man kneeling at his feet; and the merrymaking over the fatted calf.

Although Philip had come primarily for Art alone, it was impossible for him to ignore the stories the artists had chosen as foundations for their compositions. In "St. George and the Dragon," for instance, who could help making up little bits of the story that had brought the bold St. George to the mouth of the rocky den where lay that very stupid and malicious monster with one cruel paw holding a victim at its feet?

Even that brilliant piece of coloring, "The Birth of the Pearl," required the story-telling faculty to account for the swift bubbling plunge of the diver who opens the iridescent shell beneath which the Pearl Maiden is sleeping. A story nearly as good as the "Sleeping Beauty" was told in those gem-like colors.

A VIEW OF THE FINE ARTS BUILDING FROM NEAR THE NEW YORK STATE BUILDING.

Of more direct interest to the boy was "A Singing Lesson in a Public School in Paris"; and Philip gladly would have spent much time in reading the little touches of character that made each boy in the crowded picture so interesting a figure. But he knew that he must slight many pictures in order to give any time at all to those which held him before them by making him forget everything else; so he went on to the next gallery. He was first delighted by "The Bath of the Regiment," a barrack-scene showing the members of a regiment passing one by one in front of a hose in full play: the spattering water, the wet floor, the shining skins of the soldiers were wonderfully rendered considering the difficulty of painting the details from nature.

Another striking picture was the portrait of Pope Leo XIII. Philip recalled having read that the Pope had never before granted any artist a sitting; but that M. Chartran, being granted an audience, made a sketch that so pleased the Pope as to gain for the artist permission to paint this wonderful picture. The expression of the face was purely intellectual and refined, and Philip felt sure the picture would never be regarded as other than a masterpiece. There were two small portraits by Weerts that were worthy to be ranked with this larger one. Two others, landscapes, also claimed attention, one a dainty bit of bright color by Gagliardini—a Moorish scene; and the other, by Lhermitte, "Harvesters at Rest," showing peasants in the field. The only other picture that Philip marked upon his catalogue was a group of children in an arm-chair, by G. Dubufe, *fils*.

IN FRONT OF THE FINE ARTS BUILDING.

Speaking of Philip's catalogue, it is well to say that he bought two. The first was so arranged that after walking through one room with it he returned, and paid three times as much for the second. The more expensive catalogue numbered the pictures as they were hung upon the walls, and he could find each picture at once—a matter worth considering when he knew he could not see a third of the rooms in each of which were many masterpieces.

BOY WITH A DOVE.
Carving in ivory by Asahi Hatsu.

Entering another gallery, Philip drew a line of approval against "A First Proof," by Mathey—a printer examining the first impression from a plate; a similar line was awarded to "The Struggle for Life"—a marine showing a long line of men trying to draw a fishing-boat through the surf to safety. Others he marked were a soft evening effect by Zuber, and, in the next gallery, "The Virgin's Thread," that lovely painting by Lucas, where the birds are pulling at the thread while the virgin is sleeping in her chair beside the wheel. A picture of a boar at bay, while the hounds snarl, and whine but hesitate to come to close quarters, and a "Strike" picture, also compelled him to halt and to enter.

"LITTLE NELL," FROM A GROUP "DICKENS AND LITTLE NELL," BY F. EDWIN ELWELL.

But he felt as Ali Baba must have felt in the treasure-cavern—dazzled, longing to take all he could, but hurried and ill at ease. It is easy for an arm-chair philosopher to advise patience and coolness; to say, "Select a little, and see it thoroughly"; but to be a visitor at the greatest of World's Fairs is quite another matter, and in the Art Galleries you can never tell what you are losing.

They issued in Chicago several useful little handbooks to the Fair. "The Time-Saver," "The Nutshell Guide," "Gems of the Fair," "What to See and How to Find It," were some of them, and by reading these one could be fairly sure of not overlooking many "best" things in the trade exhibits. But in the Art Galleries such books can be of little use. The pictures Philip looked at pleased him for various reasons. Some were by consummate colorists; some told a pleasing tale; some preached a little sermon; some were amusing, and others played upon deeper chords. Now, as to these no two boys or men would feel just alike; and you can no more let another pick out your pictures than you can let a stranger order your meals.

As Philip was standing in one of the galleries an old man said slyly:

"No awards here."

"Is that so?" asked Philip in surprise.

"Yes," said the old man; "the French found the Germans were beating them, and so they quit!" And the old man disappeared in the crowd, chuckling to himself, and seeming to take more interest in this bit of gossip than in the pictures.

Philip went on through two rooms containing pastels and water-colors; he meant to skip them entirely. It was not that he undervalued these mediums, but he felt he had to draw the line somewhere (as in the old story of the man who didn't invite his parents to his wedding); and the oil-paintings were more numerous.

But he was compelled to look at three pictures by Boutet-de-Monvel because they were just what he liked, at one by Maurice Eliot, and at some hunting-dogs resting by a river, painted by Oliver de Penne. He made up for this pause by skipping two large collections of miniatures, etchings, and medals, and began to go around the room known as "Gallery 45."

Here he found two pictures that have caused much controversy—one showing the Crucifixion as upon Montmartre, Paris, and the other representing Christ as sitting at table in a modern drawing-room.

A PART OF THE GREAT PAINTING, "THE FLAGELLANTS," BY CARL MARR.

"THE MOTHER." PAINTED BY ALICE D. KELLOGG.

Philip didn't pretend to say whether there was a great moral lesson conveyed by this strange device; but he felt that the pictures were as unpleasant as they were powerful; and that of the Crucifixion was certainly full of intense feeling rendered by the hand of a master.

But it is useless to quote from the catalogue as Philip marked it; for himself the markings were useful, and helped him to fix his attention upon certain pictures; but unless all of the pictures are at hand, comparison and comment can have little value.

As the boy went through the galleries, he felt a strong sense of gratitude to the hundreds of skilful, keen-sighted men who had studied nature and mankind until they could show him in an instant's glance just how things were and are the whole world around.

A FELLOW-CRITIC.

From the French exhibit he passed to that of American artists; and again he found reason to be proud of his young country. Perhaps it was as well that the French and American exhibits were distinctly labeled, for there was not such a difference as there might have been. But if America showed that she had taken lessons abroad, she at least gave her teachers no reason to be ashamed; and here and there was seen a touch of true individuality promising a distinction and a difference in the future.

Julian Story's painting of "Mlle. de Sombreuil" and Carl Marr's great "Flagellants" were two history-lessons which no boy could forget; and the second of these artists, in another painting where bits of real sunshine come flickering through a screen of green leaves, showed that he could paint pictures, even without telling historical incidents. Philip went close to this picture to see just how that shining sunshine was done; but he was surprised to find nothing to explain the brightness of those shining spots except a little dull ocher paint gradually lightening to white.

After he had seen, in the next room, Douglas Volk's "Puritan Girl" and Hovenden's "Breaking Home Ties," he became a little depressed; but was cheered up by Toby Rosenthal's comedy, "A Dancing-Lesson of our Grandmother's."

When he went outside to sit upon the steps for a moment's rest, he began to understand Sir Isaac Newton's simile about picking up a few shells on the shore; for he saw that he had been several hours in the Art Building, and had seen hastily only a part of one wing of the great storehouse. He hurried back, rushed blindly through several rooms, and tried to take a small piece out of Great Britain's display. Again he was caught here and drawn there by the magic brush of one artist after another, and had to confess that he must raise the siege and hope for another day. He walked down the steps with a sense of injury and loss, which remained with him until the outdoor air and the breeze from the lake had restored his good humor.

THE GRANDMOTHER OF THE SWEDISH ARTIST ZORN.[2]
From the original carving in birch-wood (six inches high) by Zorn.

He concluded to walk home, and made his way to the path that ran along the lake-shore. Philip found his muscles a little sore, and seeing a vacant bench, sat down upon it. In a few moments he saw a group of young men pointing out upon the lake. He looked in the direction they indicated, and to his amazement made out the "Santa Maria" under full sail and as independent as any steamer of them all. Philip felt as if he might be an Indian viewing the first coming of the caravel, and wished sincerely that he were aboard, so that he might shut his eyes and imagine he heard that first cry, "Land! Land!"

He was delighted with the chance that had brought him the sight of the caravel at sea, and wondered what nabob of the Fair was cruising about as if he were Christopher Columbus himself.

Resuming his walk, he went through one or two of the buildings in order to get out of the sun (which beat down quite fiercely, considering how late in the year it was). In the Liberal Arts Building it seemed that only frail pieces of plate-glass protected the rich treasures of gold and silver arranged in the jewelers' show-windows, and Harry wondered whether a modern Dick Turpin, or Blackbeard the Pirate, could not, by dash and nerve, succeed in carrying away enough plunder to support him forever after in some reputable line of business. The pirate, he thought, would have the better chance; for he might rush to the shore, where his trusty crew were awaiting him in the long-boat, be rowed to his stealthy black vessel, hoist sail, and away with all that Tiffany and the Gorham Company had left out of their safes!

Then what a scurrying to and fro! Sailors and soldiers, losing their presence of mind, would dash up to the conning-tower of the battle-ship "Illinois" and press the dummy electric buttons, wondering why the engineer didn't get up all steam and put on full speed at once. Others would leap into the "Viking" and start to row with the long sweeps, forgetting that there were only shields aboard.

Philip was amused at this odd fancy, and resolved to ask Harry to make a sketch of the pursuit. Meanwhile he made his way home, keeping in the porticos where it was shady, and avoiding the clayey mud left by the previous day's rain.

"I'd rather," he told Harry that night, "miss some of the regular exhibits, if I've got to take the Fair in samples; when it comes to missing pictures, you never know what you've lost."

The next morning Mr. Douglass, who was reading the "Chicago Tribune," burst out laughing. "Philip," said he, "here is part of an account of the cruise of the 'Santa Maria'—the cruise upon which you saw her." And, interrupted by the boys' occasional chuckles, he read aloud as follows:

> The old caravel stood out on the waves, queer-looking as compared with modern craft, but full of grace and beauty. When the big square sail was first spread, it took the wind nicely, rounded the pier, and sailed off to the northwest in splendid style.
>
> But when the passengers wanted to turn the caravel there was trouble. Had they continued in a straight line to Michigan all would have been well, but they knew not how to sail the "Santa Maria." The craft wobbled. The choppy waves tossed it. Though it had braved storms on the Atlantic, it trembled, and its sails became disorganized by the turbulence of the white-topped waves of Lake Michigan.

"It will not sail close to the wind," said a passenger who claimed to have been out on the lake before.

"Better slow her up," suggested passenger Millet; "we're headed for the Forty-third street reef."

"Who ever heard of a reef on a street?" petulantly returned Sailing-master Hunt.

On flew the caravel until the cheeks of the passengers turned pale, and they pleaded with the captain to turn it about. Its huge hulk was finally swerved just as it scraped the reef. Away it shot again out northwest, more unruly than before.

An hour or so went by. The "Santa Maria" still sped on toward the Michigan fruit-fields. The passengers became hungry. They wanted to go home. A turn about of the caravel was finally made. It shot away toward the Van Buren street pier.

"Land her!" "Land her!" "Ground her!" cried the passengers.

With care the caravel was brought up near enough to the pier to let off the passengers, and the craft was anchored for the night.

Then Mr. Armour said to Millet: "No wonder Columbus discovered America."

"Why?" inquired the latter.

"Because a man could discover anything in such a craft as the 'Santa Maria.' There's no telling what direction it would carry him. The discovery of America was a 'scratch.'"

[2] With regard to the little bust of his grandmother, carved in birch-wood, Mr. Zorn says: "I have painted my grandmother a great many times, and the

pictures have always been sold, so I made this little carving as something to keep. From beginning to end it was carved from nature and with carvers' tools. My grandmother," he adds, "is very picturesque"; but this we do not need to be told, nor that there were probably other reasons why her grandson wished to have a portrait of her; nor again, that this bust probably is a portrait in the fullest, exactest sense of the word. It is a delightful thing in subject as in execution. Every detail of the sweet, strong old peasant face is lovingly rendered, and yet one thinks most not of details or even of features, but of the soul behind them.—*"The Century" for August, 1893.*

CHAPTER XI

Going after Letters — The Agricultural Building — Machinery Hall — Lunch at the Hotel — Harry's Proposal — Buffalo Bill's Great Show.

ON Saturday, Philip had heard that for five dollars he could secure permission to use his kodak for a week, and by going to the office of the official photographer on that day and paying the necessary amount, he was able to dismiss from his mind any anxiety about carrying his camera. So on Monday the two boys and Mr. Douglass entered the grounds, fully equipped with note-book, sketch-book, and camera.

Hitherto Philip had been asked but once to exhibit the license, but this time he was challenged by one of the ticket-takers, who shouted to another, "Hi, Jack, here's a kodak!" But, as it turned out, neither ticket-taker cared to examine the card, and Philip merely waved it, saying, "It's all right."

The day was too rainy to risk taking snap-shots, and Philip carried the camera during the forenoon only, and was glad to leave it behind at the hotel when he returned to lunch.

They had down on the list for this day a trip to Chicago; but had asked to have the date of their tickets for the coach changed when they saw the sky was gloomy and overcast. Instead of going into the city, therefore, they resolved to give their morning to the Agricultural and the Machinery Buildings. They walked first to the Manufactures Building to get letters, and took a launch back again. While waiting for the boat they had some conversation with the man at the landing, and were surprised to learn that each of the launches cost more than three thousand dollars—the high price being paid mainly for the machinery.

THE AGRICULTURAL BUILDING—TOWARD EVENING.

Landing at the Agricultural Building, they were glad to escape the rain—a thunderstorm—by entering at the main door. The exhibits seemed to be arranged according to nationalities, the first one they came upon being that of Porto Rico; and the boys were really surprised, upon exploring their minds, to find out how little they knew about Porto Rico. Mr. Douglass knew a little more: he told them it was an island—one of the Greater Antilles—and belonged to Spain; but there he came to a sudden stop, and directed the boys' attention to a miniature fort in which bottles of wine served as guns. Having to that extent improved their knowledge of Porto Rico, they moved on a few steps, and seemed to have walked into a cigar-box. The odor was explained when they saw before them Cuba's display, which was not unlike that of a prosperous tobacconist. British Guiana did not repel them, though a woman cried out, "Oh, alligators and snakes!" as she turned hastily away. She was followed by two more of the less timid sort, one of whom said resolutely, "Come in. I want to see this alligator. I never saw one in my life"; to which her companion replied, "Well, gaze on him; there he is!"

"You might think, boys," said Mr. Douglass, as the boys smiled at this dialogue, "that such people got no good from coming to the Fair. But I think such a conclusion would be a mistake. The foolish chatter we hear has little to do with what people are really thinking. They cannot help picking up clearer ideas of the world and its inhabitants as they go through these buildings. Where one sees fruits and grains, it means that this or the other country has orchards and farms. We thus get rid of many a foolish mental picture. We cease to imagine that all the Chinese are continually flying kites and smoking opium, or that all Spaniards are eternally strumming guitars in the sunshine. You may not think you have such foolish ideas, but you will probably find yourself entertaining notions quite as absurd. I only say this because we hear so much trivial chatter that you might be misled by it."

"Well, Mr. Douglass," Harry answered, "I have seen plenty of men, and women too, who are taking the Fair almost too seriously. And even the most foolish must find a great deal that makes him think. I know I do. Now, for instance, look at that figure"; and Harry pointed to the model of a negro workman that made part of the exhibit labeled "British Guiana."

"I saw him," said Mr. Douglass, "and I noticed how his leather sandals have absurd twirls and coils of leather thongs about them. The rest of his dress is very ordinary."

"Those are just what I mean," said Harry. "I said to myself, at first glance, that those twisted rolls of leather were silly ornaments, and showed that the man was a savage in civilized clothes. Then I wondered whether they hadn't some use, and—"

AGRICULTURAL BUILDING, NORTH FRONT, SEEN FROM THE GRAND BASIN.

"I see," said Mr. Douglass, interrupting.

"Well, I don't," Philip declared.

"Suppose he should break a sandal-string," said Harry, eagerly; "don't you understand that he could just untwist one of those coils, just as a violin-player unwinds a little more of his E string?"

"Yes, of course," Philip said; "and that is the most convenient way for him to carry the strings."

"I have little doubt that the coils came from that necessity for mending," Mr. Douglass remarked; "but probably the dandies exaggerated the coils. This idea of yours, Harry, reminds me of an article by Remington, the artist. It was written to show that good sense dictated the whole costume of the Western cowboy. I kept it, and will show it to you."

Liberia displayed various native products, and fine works in metal and straw and leather, but the party did not see anything to warrant a long stay; Mexico had so arranged her exhibits that they reminded one of a grocery kept by a neat but eccentric grocer; but wherever the flag of Japan was displayed, the boys never grudged time for examination. That artistic little nation can always teach a lesson to natives of the young Occident. Even in their display of food-stuffs, the boys found the pickle-jars, saké-kegs, and some boxes worth looking at. In fact, Harry was so pleased with these artistic groceries that his sketch-book came out at once. The pickle-jar was covered with white paper draped in graceful lines and tied down with a twisted purple cord and tassels! The saké-keg and the box also showed the same wish to please the eye and satisfy the needs of each article; and as for some larger jars, they were dressed as richly as a ball-room belle. They left the domains of the white flag and red disk with some instruction in the art of "framing" groceries.

ONE OF THE PANELS ("SUMMER")
IN THE PORTICO OF THE
AGRICULTURAL BUILDING. PAINTED
BY GEORGE W. MAYNARD.

Whenever the party first entered one of the exhibition buildings, they examined the earlier booths somewhat carefully; but a sense of losing time soon made them hurry on. So it was then. They walked on by many a fine arrangement of food-products—notably those of the Western grain-producing States. They admired the taste and skill that had utilized glass tubes full of grains as columns, and corn on the cob as a building material. Mr. Douglass said that much foolish criticism had been evoked by these booths, but that a sight of the structures themselves called for approval rather than fault-finding. They particularly admired the displays of Wisconsin, Nebraska,

Iowa, and Ohio, in which both the general effect and the bits of color decoration showed good taste and much constructive skill.

"I shouldn't be surprised," said Harry, "to come upon a Greek temple built exclusively of old shoes."

Here they were stopped by a bit of fun. A bright-faced young woman was throwing little tin forks out among the crowd; these tin forks advertised a brand of sardines, and were made in the shape of a little fish, the tail reaching to the tines of the fork. Picking up the forks, the boys naturally went to see the exhibit, and were invited to take a sardine, free, from an open box. They declined, but others were not so lucky. One old man eagerly plunged his fork into the box only to discover that the fish were painted tin. He fled into the crowd while the bystanders laughed at him. This device certainly attracted plenty of attention, but whether it was wise was doubtful.

GREAT CENTRAL PORCH OF AGRICULTURAL BUILDING.

They finished the aisle they were in, and crossed to another, which they walked down, having gone up the first.

In the Greek exhibit they saw some tobacco labeled as from Thermopylæ, which at the moment seemed incongruous; but reflection showed that Thermopylæ must be something beside a battle-field. Louisiana had built herself an Egyptian temple of sugar-cane, and again Harry made a sketch, for he found the effect very pleasing. Passing a number of other booths, they at last came to the agricultural implements, and found that there was more to know than shovel, spade, and hoe, or even plow and harrow. They frankly confessed ignorance of the mechanism and purpose of most of the nickel-plated apparatus, and concluded that in their present state of ignorance time spent here would be wasted. They did smile, however, at seeing a harvesting-machine labeled: "The judges ordered this harvester to be tried in a field of standing grain. It is a little disfigured, but still in the ring."

A sign revealing the location of the "Sandwich Manufacturing Co." somehow reminded them that they must see something of Machinery Hall before lunch, and they started toward that building, passing on their way a "prairie-breaking plow"—a rude but enormous implement that had been used with a team of six or eight oxen in first turning up the new Western soil.

As they were coming out, they paused, even in the rain, to admire the fine proportions of the Agricultural Building; its dignified portico, the fine groups and single statues that adorned its principal features,—such as Martiny's "Abundance," for example, and the signs of the zodiac, and the great corridors that unite Agricultural and Machinery Halls.

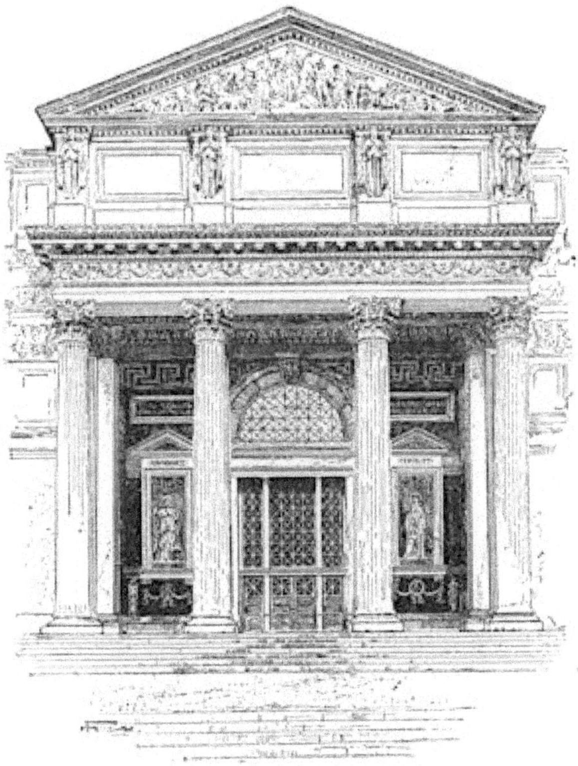

PORTICO OF THE
AGRICULTURAL BUILDING.

Upon entering Machinery Hall, and finding that they could not give anything like adequate time to it, they went at once to the gallery and waited for the traveling-crane. There are three of these, each originally used for putting in place the heavy exhibits; they run upon great girders supported from the floor upon uprights similar to those upholding an elevated railroad. Moved by electricity, they traverse the whole length of the building and then return, carrying passengers twenty or thirty feet above the crowded floors and at an excellent height to permit of overlooking the show.

They had to wait a little while, but soon the great floating beam of iron came against the edge of the gallery, almost as lightly as a bit of thistle-down, and they entered at one end and sat down upon chairs ranged along the front edge. The crane carried them to the other end and back again for ten cents, and without effort they had at least a glance at all the exhibits in that part of the Hall—thus obtaining, no doubt, a better idea of what there was in the building than could have been secured in a long walk below.

In order to show how bewildering were the displays, here is a list that Philip made while Waiting for the crane to move. It shows only what he could easily make out from the extreme end of the hall. There were machines relating to hot baths, candy, lubrication, ice-cream, smokeless furnaces, rock-drills, galvanizing, window-washing, and baking.

THE CONNECTING SCREEN OF CORRIDORS BETWEEN THE MACHINERY AND AGRICULTURAL BUILDINGS.

They found the ride cooling and breezy, and saw enough to greatly interest them on all parts of the floor. The enormous printing-presses were especially "impressive," as Harry put it, and one press was printing colored illustrations of the World's Fair buildings. Besides, they noticed many looms, sewing-machines, a spool-cotton exhibit, dyeing works, glove-making- and washing-machines—each something novel or interesting. They attempted to see all they could, and keep eyes and brains active; but Harry said it reminded him of the small darkey who "slipped back two steps for every one he took forward"; for they missed two exhibits by pausing to examine any one. They had meant to take a ride upon the other crane; but when they saw there were three, they agreed, as usual, to be content with a half-seen show, and departed from the grounds, going back to their hotel for lunch.

The dining-room, so crowded at breakfast- and dinner-time, was almost deserted at noon; and they found they could talk over their plans with perfect freedom.

Mr. Douglass and Philip made several proposals: the Art Galleries; another visit to Machinery Hall; more State buildings; the Anthropological Building—an inexhaustible resource. But Harry shook his head at each suggestion, until at last Philip said:

"It's plain that you have a plan of your own, and I've a good mind to veto it anyway. What is it?"

FIGURE IN WINDOW-FRAME OF MACHINERY HALL.

"I have wasted my time very patiently with you this morning," Harry said gravely, "because I suppose we ought to 'do' the Fair. But I remember that the English poet said, 'The correct thing for man to study is man.' See? Now, we have been looking at staff and iron and steel and corn and wheat and bottles and strings and other precious metals all these hours. I have gone through it, though the buzzing and rattling and thumping and worrying were decidedly unpleasant. Now I want to study man. There is near this hotel, I have learned by careful study of bill-posters' literature, a gentleman who was a member of the legislature, etc., etc.,—but who is known among us boys by the name of Bison William."

"I have heard of him," said Mr. Douglass, with a grave face.

"Who has not?" said Harry, enthusiastically.

"He is now conducting an educational exhibit near here, where one may see various nations at their sports and pastimes. And, gentlemen of the jury, what I say is: Let the machinery whirl, and let us devote ourselves to the Wild West Show. What do you say?"

"I'd like to go," said Philip; "but I wish it was a better day for taking pictures."

"I'm willing," said Mr. Douglass. "I saw the show some years ago in New York, and it was well worth seeing. I am not sure that a whole day of systematic sight-seeing at the Fair is not a little too much when one is busy at it for a week or two at a time. Where is it?"

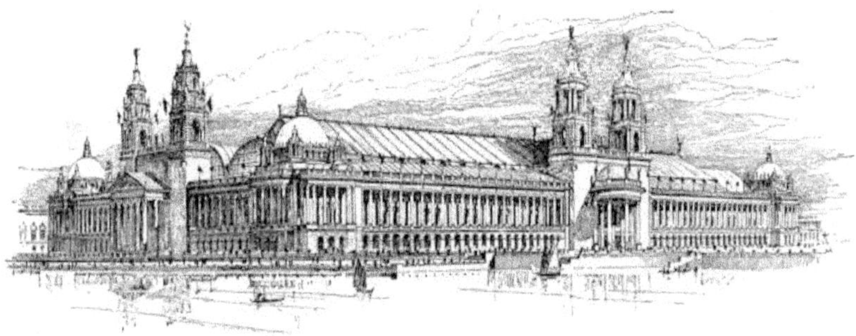

MACHINERY HALL.

A SUGGESTION OF THE "WILD WEST."
Remington's Famous Picture, "A Bucking Bronco."

"Just around the corner," Harry answered. "And Phinney says it is twice as good as it used to be."

A short walk from the hotel brought them to the grounds, a great square open space around which were seats like those upon base-ball grounds. They bought tickets for the grand stand, and gazed expectant upon a sea of mud. The sign said "Rain or Shine," and rain it was: no drizzle, but a pelting downpour that roared upon the roofs overhead. Boys walked to and fro, one crying, "Sour crystallized lemonade-drops—souvenir in every package," and the other, "Peanuts!—are five cents!"

AN ABORIGINAL.

The rain plashed in the puddles upon the arena, and the boys were not sorry; it was a new sensation to see a performance in the rain. A band played loud enough to be heard nearly to the Rocky Mountains, a man in a very broad-brimmed felt hat mounted a rostrum imitating a boulder, put on a rubber coat, and, when the band was hushed, began a speech at the top of his lungs,—so loud that he hadn't breath for more than a word or two at a time. He said, "Ladies—and—gentlemen:—From time—to—time, I shall—announce—the nature—of the—display," and so on. One seldom hears so forcible an oration.

He announced one by one the bands of Indians, their chiefs, the white men, their captains or leaders, and each of the items upon the program. But his shouts can be omitted with the assurance that he did his level best. One example will be enough.

"The Arapahoes!"

A gate is unbarred, yells break through, and helter-skelter come a troop of almost naked savages painted and bedecked, riding their ponies at a run. They draw up before the grand stand.

"Their chief!"

A single Indian comes flying across the field lashing his running pony, and draws up before his band.

Then, in order, come other tribes until a motley, bright-colored rank of mounted warriors are ranged all along the front of the field. Then French cavalry ride in with similar heralding, except that the color-bearer is announced separately and the band plays the Marseillaise. German lancers follow to the tune of "Die Wacht am Rhine," and after them, Mexicans, American cowboys, British Lancers, and Cossacks perched on high saddles. The Indians are holding their shields above their heads to protect themselves from the rain. Now Arabs come, and two women riders; an old guide, gray-bearded and dressed in fringed buckskins; United States cavalrymen, riding upon gray horses; and at last, cheered even more than the Stars and Stripes, there gallops to the head of that great array an honorable gentleman, of whom Harry remarks: "That is Biffalo Bull himself—and a fine-looking man he is!"

At a signal from the scout the whole cavalcade springs into life and rapid motion. The plain is dotted with horsemen dressed in gay uniforms; and just then the sun breaks out to brighten the scene, and a rainbow is seen above the right-hand portion of the grounds as the riders follow one another out. It was certainly a brilliant and cheerful pageant.

A well-known markswoman runs over the liquid mud, making swimming motions with her arms, and taking up a gun breaks clay pigeons and glass balls as fast as they can be supplied by the attendants. Fancy shooting follows, and, making a miss, the woman walks around the table where the guns are resting. This whimsical performance makes the people laugh.

Several usual features follow. A race between riders of different nations; the "pony express," an exhibition of rapid shifting from one horse to another; an emigrant-train attacked by Indians, but saved by the blank cartridges of the Hon. Mr. Cody and his rough-riding friends; and then come Syrians and Arabians in wonderful feats of balancing, juggling, and pyramid-grouping. In this last act one of the men supported nine others in the air—a weight of perhaps twelve hundred pounds.

"And yet," Harry remarked, "some men find it hard to support a small family."

Always interesting, the thick mud made the show funny as well. It was hard for men and horses to secure a foothold: Syrian acrobats stopped to wash their muddy hands in almost equally muddy water; some of the fierce horses were compelled to drop almost into a walk instead of running madly across the arena; when a marksman wished to lie down in order to shoot from that position, it required careful search to find firm ground for his blanket; the men who built themselves into pyramids bedaubed one another until their dresses were mud-color instead of crimson; and all through the long, delightful program the sticky mud took a prominent part in amusing the spectators.

When "Old John Nelson" rode up near where the boys sat, and delivered the mail from the old original Deadwood coach, he hurled it off with the regulation speech, "Here's the Deadwood mail," and then added, winking to Harry, "A little damp, too; but never mind!" The same genial old guide, who was lying lazily across the coach roof, raised himself coolly as the scouts cried, "Indians! Indians!" and again grinning at the boys, remarked in a low tone, "Going to be Indians, eh? Then I'll get up!"

A COW-BOY.

This by-play delighted the boys; but best of all was "Custer's Last Charge."

First came the Indians, and encamped far away across the plain. A scout followed; discovered them with plenty of gestures to let the audience into the secret; reconnoitered them over the imitation rock; rode off to tell "Custer" and his staff—mainly buglers—of the great find; brought back the general, who gazed meaningly at the red villains through a warlike night-glass, and then all the white men retired for reinforcements.

Coming back, the cavalrymen charged fiercely on the Indians, fired off several dollars' worth of gunpowder, and disappeared behind a curtain. Mournful music indicated the terrible fate of the cavalrymen.

During the whole afternoon the boys sat beside a boy from Chicago who told them many particulars about the show and the riders. He said he had seen the performance four or five times, but seemed nevertheless to enjoy it. Harry learned that the young Chicagoan sometimes came to New York city, and gave the boy his address, inviting him to call.

It began to rain again as they went home, but it was only a short distance to the hotel, and they went straight to that goal in spite of a most pressing invitation to "Take supper here now for twenty-five cents, and go home by the light of the moon!"

Harry was rather silent on the way home, but showed the course of his thoughts by remarking: "I think perhaps I will give up being anything too civilized; I'm going to ask my father to buy me a ranch far out West."

"I wonder," said Mr. Douglass, "whether the young Indians who come to the Fair with the Indian schools ever go to see the Wild West Show?"

A CHICAGO STREET.

FORT DEARBORN. (CHICAGO, 1804–1816.)

CHAPTER XII

The Tally-ho — How it dashed along — The Parks along the Lake — Chicago — The Auditorium and other Sky-dwellers — The Whaleback.

MEMORIAL BUILDING, ON THE SITE WHERE THE GREAT FIRE STARTED.

ON Tuesday morning the party hurried through their breakfast in order to catch the tally-ho which was to pause in its mad career to pick up passengers from their hotel. Although it was a cloudy morning, threatening rain, they did not like to postpone this trip again. Consequently ten o'clock, the hour set, beheld them "all agog to dash through thick and thin" like John Gilpin.

Presently something drew up at the door. It was not what would be called by the critical a tally-ho. It was not even a coach. It was on wheels, it had seats here and there, and four animals dragged it. Baron Munchausen once had his horse cut close off by the fall of a portcullis. If the same accident had befallen a tally-ho, and it had been then spliced to the end of a park wagon, the resulting vehicle would have been not unlike the wagon which presented itself at the door.

"Is this it?" asked Mr. Douglass, dubiously presenting his ticket.

"This," said the man (he was hardly yet a voter), "is it. Yes, sir. The tally-ho, sir."

"Well," remarked Mr. Douglass, turning to the boys, "what do you say?"

"We'd better go," said Harry. "It's all arranged; and the wagon looks comfortable anyway. Don't you say so, Phil?"

"Yes," said Philip. "It's no tally-ho, but I don't know as that makes any great difference. It has wheels, and—horses," after a pause.

Having taken outside seats, they climbed up on the wheel-hub and two steps, and were soon perched some ten feet above the ground ready to start. Just as they settled themselves in their places, a policeman came to the curb and spoke warningly to the driver, who said, "I can't help it," and gathered up the reins.

DRIVEWAYS OF THE GRAND BOULEVARD.

Mr. Douglass, who was not used to fast riding, made up his mind that their lofty seat might be a risky place to sit, and was gratified to find a stout rail at the back of the seat, which afforded an excellent place to hold on. Harry, too, concluded that they would soon be tearing at breakneck speed through the crowded streets of the city, and began to think he had been unjust to the "tally-ho."

"We're off!" said Philip, as the horses heaved at the traces and the wagon changed its place leisurely: At a slow walk they drew the wagon around the corner and stopped at another hotel. A man who seemed to be in charge alighted and entered the door. That was the last seen of him for a considerable period. Queries to the driver were smiled away. They waited and waited. Nothing happened. After their patience was gone, the missing man came back, and the coach floated on.

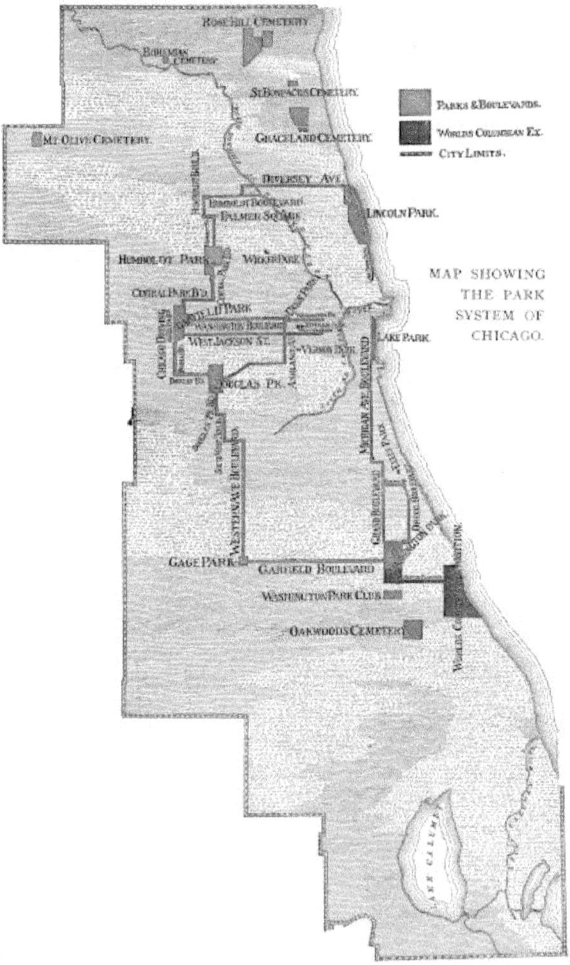

MAP SHOWING THE PARK
SYSTEM OF CHICAGO.

"Now we're started!" said Mr. Douglass, with an expression of relief. But the coach rounded a corner in a leisurely manner, and drew up at another hotel. Again the man disappeared, and the waiting was repeated.

"This is not a tally-ho," said Harry, "it is a tarry-whoa"; and so it proved. Even after the man was again at hand, the old coach went no faster than the slowest of jog-trots. And at the same dolorous gait they loitered along on Woodlawn Avenue, a straight street beautifully paved, and fit to be a blessing to bicyclers. They were as long in passing a given point as was possible. Every vehicle went by them except children's carriages with nurses; wagons of heavy iron-castings, dirt-carts, streetcars—until one man remarked jocosely that he was afraid a funeral might come up behind and run over them.

Then Harry remembered the policeman who spoke to the driver just as they were starting, and a light dawned upon that mystery.

"You remember that 'cop' who talked to our driver?" he asked Philip.

"Yes," said Philip; "I thought he was warning him against reckless driving."

"So did I," said Harry, laughing. "But I'm sure now that he was saying a word for the poor horses. Why, those Fifth Avenue stage-horses they make such fun of in New York are Arabian coursers compared to these! See them creep!"

VIEW ON STATE STREET, LOOKING NORTHWARD FROM MADISON STREET.

They passed some gray stone buildings on the way to the business part of the city, and the driver said they were the Chicago University—a statement they accepted at the time, but doubted when they became better acquainted with the driver's acquirements as a guide. Another great establishment they saw was an old field crowded with tents and labeled "Camp Jackson." A sign upon its rainbow-tinted fence informed the public that board in that field and under those tents was two dollars a week and thirty-five cents a day.

"It's a comfort," said Harry, "to reflect that all these places, rough as they are, mean to offer Fair accommodations."

At another time this weak pun would not have been noticed, but upon that weary, slow ride anything was a relief: when the horses stopped to drink, it was an event; when a new passenger got on (one did), the excitement was intense. But nothing hastened the wagon. It meant to get to Chicago if it took all day; and after awhile they did begin to see buildings more closely set, and then they entered a beautiful park. The driver said it was Washington Park, and on consulting a map afterward, the boys made up their minds that he had guessed right—there were some things the driver knew.

The park was flat as a board, as is all the country for miles around; but as the ground was mainly given up to beautiful green lawns extending as far as one could see, the effect was excellent, and marred only by some very florid designs laid out in colored plants. One of these designs formed a sun-dial, called "Sol's Clock"; another showed a few bars of "Hail Columbia."

THE POST-OFFICE. THE CITY HALL, CHICAGO.
(From photographic prints. By permission of C. Ropp & Sons, Chicago.)

Even Mr. Douglass had now given up his visions of dashing along to the sound of a "yard of brass," and so far from being at all nervous, would not have been afraid to stand upright in any part of the coach. He kept thinking of a parody upon Shakspere's description of the school-boy: "A tarry-whoa, creeping like snail, unwillingly to Chicago." By this time they were in Michigan Avenue,—a thoroughfare with beautiful grass plots along the street, but houses that did not please an architect who was also on top of the coach. He declared all but a few of the houses to be fussy and tiresome; and the boys noticed that those he commended were plain and simple in their outlines, and little decorated.

At Twenty-second Street, they saw the Chicago street-cars, and found that they ran in trains of three coupled together, an arrangement of which they heartily approved. As they passed a baker's cart, a small boy leaned out and whipped the horses of their coach; whereupon several of the passengers thanked him warmly, even though his efforts produced, no result. Still, in time they did reach the city, and recognized the lofty Auditorium, an enormous pile of stone, so many stories high that the boys lost count in attempting to reckon them. Soon after they admired the Art Institute, "a broad and low building of impressive design." They also saw the foundations being laid for another great building, and remembered having read in *St. Nicholas* that these heavy structures could be supported only upon artificial foundations, such as long piles driven deep into the soft ground. The Masonic Temple was also seen as they passed through the busy part of the city.

There was a smoky smell in the air, and their first impression was of being down-town in Broadway, New York, when a great fire was raging, filling the air with smoke. Possibly the smoke was worse than usual, for rain was falling at intervals and the air was heavy.

HOUSE OF JOHN KINZIE,
THE FIRST WHITE SETTLER.

None of them talked much, for the slow drive was anything but enlivening. They went along Lake Street for some time, and then wandered on until they drew up at the Waterworks. Here, despite the protests of the passengers, there was a halt of five minutes, and some got out and went in to see the machinery. When all were on board again, the scenery slowly changed, and they found out that they were in motion once more. But as they had reached the Lake Drive,—a beautiful boulevard, and one of the system of drives that encircles the city, connecting Chicago's great parks into a ring of pleasure-grounds,—the slow driving was not so irritating. They saw Mr. Potter Palmer's castellated mansion fronting the lake, and passing other fine dwellings, reached Lincoln Park.

Against the sky, in silhouette, appeared the statue of General Grant, an impressive feature of the park, and they were sorry that their route did not bring them within view of the even finer Lincoln statue, of which they had seen many pictures. Looking forward along the drive, they saw a dark point of land along the horizon beyond the lake, and were told by the rather taciturn driver that it was the city of Milwaukee, which information surprised them quite as much as if he had said it was Bagdad. "Traveling certainly makes one modest," said Mr. Douglass, who doubted the driver's statement. "I had no thought of seeing Milwaukee upon this drive."

THE AUDITORIUM.
MICHIGAN AVENUE
AND CONGRESS STREET.
(From a photograph
by J. W. Taylor.)

THE ART INSTITUTE.
MICHIGAN AVENUE.
(From a photographic print.
By permission of
C. Ropp & Sons, Chicago.)

THE WOMAN'S TEMPLE.
LA SALLE AND MONROE
STREETS.
(From a photograph
by J. W. Taylor.)

MASONIC TEMPLE.
STATE AND RANDOLPH
STREETS.
(From a photographic print.
By permission of
C. Ropp & Sons, Chicago.)

Another statue of a man in an old-style curled wig was seen, and the driver told them it was Linn. Even the tutor had never heard of Linn, and all remained puzzled until a turn in the road showed the inscription "Linné," whereupon they recognized Linnæus. Though they hated to lose the invaluable information the driver was giving them in homeopathic doses, they were glad when the coach worked its way to the front of a park restaurant, and announcement was made that there would be a halt of an hour or more for lunch.

THE LAKE-SHORE DRIVE.

"Mr. Douglass," said Philip, "I don't know how you feel, but my feet are as cold as ice, and I'd rather get off and walk."

"Oh, let's walk!" Harry chimed in. "I'd rather ride in a canal-boat than to stay in this old coach any more."

"So would I," said Mr. Douglass. "I consider this ride a regular swindle. See here!" he went on, turning to the driver's accomplice,—a young man who rode inside,—"what is the matter with this conveyance? We've crept all the way out. Are you going any faster?"

"No, sir," answered the young man, turning State's evidence and revealing the whole secret; "the fact is, those horses—look at 'em!—are all played out. They've been going over this road for months, and they're played out."

"We have had enough of it," said the tutor, a little sharply, "and we'll walk."

"I don't blame you," the young man answered, as if he would have liked to join them.

VIEW ON MICHIGAN AVENUE, CHICAGO.

Leaving the park, they inquired how to get back into the business center of the city, and were told to take the cable-cars. These proved to differ in some ways from New York cars, and one feature seemed worthy to be copied. At the ends of each car the side seats ceased, leaving a clear floor all across the car near the door, so that those who were compelled to stand should not obstruct the middle aisle at the doorway.

"That's a good idea," Philip remarked, as he pointed out the arrangement to Harry; "for I've often noticed how people are sure to stand right in the doorway, blocking up the passage."

When they were near the end of their trip, the cars ran underground through a whitewashed tunnel, and the boys made up their minds that they were either running under the river or under the railway-tracks.

"It's about time for lunch," said Mr. Douglass, looking at his watch; and turning to a young man beside him, he asked where there was a good lunch-room. The young man recommended one, and they felt grateful to him afterward. It was a large establishment, containing several kinds of lunch-rooms. They went into the "business man's lunch-room," and had an excellently cooked meal at a fair price.

THE ROOKERY. THE BOARD
OF TRADE BUILDING.

Until it was time to take the steamer, they wandered about the city looking at the more notable buildings and enjoying the sensation of being in a strange place. The great wholesale stores were like those in parts of New York, but New York had nothing just like some of the lofty buildings of more than twenty stories. Harry said that if there were two or three streets like Broadway and running across one another, or if Broadway were cut off in sections and laid criss-cross, the result would resemble Chicago. They saw the Auditorium again, and the Chamber of Commerce building, as well as some others; but the rain was unpropitious to sight-seeing, and they soon determined to make their way toward the "Whaleback" steamer. Of course they went wrong at first, for Chicago is a puzzling place to strangers, and Harry had to ask a big policeman for directions. He was hardly old enough yet to have lost his awe of "cops," and felt relieved when the officer showed himself courteous and obliging. From what he had read of Chicago distances, Harry would not have been surprised to have been told he must "go fifteen miles south, then take a cable-car four miles west"; but their destination proved to be not so very far away.

A STREET BRIDGE ACROSS THE CHICAGO RIVER, SWUNG OPEN FOR THE PASSAGE OF BOATS.

Another cable-car rattled them down to Van Buren Street, and they found themselves, after a short walk, upon the dock awaiting the iron vessel so aptly named "Whaleback."

The boys were struck with her likeness, as she came close along the dock, to some of the dug-out canoes they had seen at the great Fair. They learned, however, from their friend the architect (whom they met again on the pier) that the boat was seaworthy, carried a large cargo, and was very fast, going even twenty-two miles an hour.

Going aboard, they found her divided into three decks, and very finely fitted up. The second deck, which was even with the top of the hull, had walks along the curved sides of the vessel; for these "tumbled home" so as to be almost level.

In the cabin, Harry found a phonograph which was advertised to sing his favorite "The Cat Came Back"; and he persuaded Mr. Douglass to try it. The tutor's face, as the song began, lost its usual quiet expression, and soon he grinned quite as broadly as the small boy Harry had sketched at the Fair. Then the boys paid another five cents, and listened to a lively song called, "Drill, ye tarriers, drill"—wherein were introduced sounds of blasting, the singing, the orders of the boss, and all the features of work upon a railway excavation.

But they wasted only a few minutes in the cabin, for the view of Chicago, as the boat steamed out, was well worth seeing. A few rays of sunshine struggled luridly through the heavy pall of dusky smoke that drifted over the city. Here and there great buildings or towers rose above the rest, but the whole effect was soft and hazy. It was a picture of the city that was sure to remain long connected in their minds with the name Chicago.

FISHING FOR PERCH FROM THE BREAKWATER, CHICAGO.

The trip was not a long one, but Harry found time to pick up acquaintance with a young man from Indiana, and the two were soon pronouncing words for each other's amusement. He found Harry's slighting of the letter R very droll, and told the New York boy that his mother had an aunt who was "a regular Yankee," and said, "Why, I could listen to her talking all day; it does sound so queer!" Harry found the Indianian's accent quite as strange, and said it reminded him of peculiarities he had noticed in the speech of Virginians.

As they approached the long pier that extended out from the Fair Grounds, Philip began to be uneasy.

"What's the row, Phil?" Harry asked, noticing that his friend was frowning rather fiercely; "are you sorry to get back?"

"The matter is this camera. I've got to take it through the grounds," Philip replied.

"I thought you had a permit for a week," said Harry.

"So I have," replied Philip ruefully; "but it is at the hotel. I took the camera along this morning, hoping that the weather would clear up so I could take something in the city; and I've been lugging it about all day without getting anything to speak of. Now here I am with no way to get to the hotel except by going through the Fair, and I haven't got a permit."

"Whew!" Harry whistled. "Two dollars out!"

But when Mr. Douglass came up, he was inclined to think there would be no trouble about the camera.

"I'll tell you what I should do," he said. "Just walk along boldly, and if any one stops you, tell them the circumstances and then face the music."

THE GREAT FIRE AT CHICAGO, OCTOBER, 1871.

Just as Philip was going through the gate, one of the ticket-takers said, "Say, is that a kodak?"

"Yes," said Philip, "it is."

"Have you a permit for it?"

"Yes," said Philip, "but it's at my hotel. It's good for a week, but I didn't bring it to-day"; and he went on to explain just how matters stood, offering to do whatever was right. "But," he said, "I'll tell you one thing—I don't want to pay two dollars just to carry this camera through the grounds on a cloudy day at five o'clock."

THE WHALEBACK.
UPPER DECK.

"I should think not!" said the man, laughing good-humoredly. "I'll find the inspector and see what he says"; and he walked out along the dock. In a few moments he came back saying, "It's all right; take it in. The inspector says he couldn't let you if it wasn't after four o'clock. You won't try for any pictures?"

"No," said Philip, much relieved; and away he went, feeling that honesty was the best policy.

Walking through the Court of Honor just at dusk, they were again delighted with the appearance of the buildings in the soft evening light. The Peristyle was especially artistic, for they saw through the columns the heavy, curling black smoke of the "Whaleback," as she set out on her return trip to the city. The gilt decorations upon graceful Machinery Hall shone brightly, and they had to stop and gaze around them with renewed delight.

THE WHALEBACK.
LOWER DECK.

"Perhaps it is just as well that these buildings are not to be permanent," Mr. Douglass remarked, as they walked on. "We like them all the better for knowing that they are, after all, mere bubbles of staff, blown to delight the eyes for a little while. The architect whom we met on the coach said to me, 'Somebody hit the nail on the head when he called these Fair buildings an architectural spree—it has been a bit of fun for the architects to show in plaster what they could do in marble; but why can't some of our cities make a similar smaller show in marble—say an ornamental building like this Peristyle, around a harbor?'"

When they asked for the keys of their rooms, Mr. Douglass received also a letter. "Ah!" he said, "here's the letter from your father, Harry. Come up into my room and we will read it." The letter was as follows:

September 21, 1893.

> DEAR MR. DOUGLASS: When I telegraphed this morning, I was afraid you would think it strange unless I promised a letter. But now I sit down to write, I feel there is little to add to my despatch. I know Mr. Farwell will arrange business

details, and that you will get safely to the Fair. I am sure you will know that I do not expect you to feed the boys on useful knowledge all the while you are in Chicago; but I should like Harry to look carefully after two things. I would like him to see the railroad exhibits, and to see the papers about Columbus. The latter is important, because there will never be so good a collection brought together again. The railroad exhibits I should like him to see, because I wish him to learn what an amount of skill and learning has gone into the modern railroad. Perhaps then the business will attract him, and I shall expect him to take it up when I must resign. As for Philip, he'll learn more about the Fair by himself than any one can teach him.

I think perhaps a fortnight should be enough to spend at Chicago; but as to that, use your own discretion. I hope that all three of you will enjoy the big show, and I'm sure you will be better Americans for having seen it.

Tell Harry that his mother and I are well, and give him our love. With warmest regards to Philip and best wishes for you all, I am your obedient servant,

HENRY BLAKE.
MR. JAMES DOUGLASS.

"That's just what I thought," said Harry. "He wishes me to get into railroading, and that is one reason he sent me here. I see one thing; I have got to go through the Convent again. I hardly looked at those old documents."

"We have a few days yet," said Mr. Douglass; "we will certainly go more carefully over those exhibits. I am glad to hear from your father, though I know his ideas well enough to have been very sure of his intention. I have still plenty of money, but I think that two weeks will be enough to give to the Exhibition. One could not exhaust it in years."

THE WHALEBACK.

GENERAL VIEW OF FISHERIES PAVILION.

CHAPTER XIII

Philip's Day — Visits the Photographic Dark-room — The Fisheries Building — The Aquaria — Fishing Methods — The Government Building — The Japanese Tea-house.

WEDNESDAY Philip had set apart on his schedule for the Fisheries Building, intending to spend any spare time at some less important places near by. He had already found that it was well to save himself what bodily fatigue he could, and so he took a rolling-chair almost as soon as he entered the grounds, from a conical tent not far from the Pennsylvania Railroad exhibit. The man who pushed the chair told Philip he was from Finland; and a few attempts to converse with him were so fruitless that Philip gave up trying.

He went first to the photographic room where he had left a roll of pictures to be developed. Then, after making the usual morning call for letters, he went on to the appointed building. On the way, the Finland guide woke up enough to show some interest in photography, asking Philip, "Do you take in colors the pictures yet?" Philip in reply gave a short account of the state of (what is called) color-photography, and the Finland guide was probably more muddled than ever.

Philip had once or twice stepped into the Fisheries Building before, but so far had never been in the east wing, where the aquaria were situated; so he selected this part as a beginning. As usual, he had brought his camera, and right at the entrance he found a good subject—a young man who was perched upon one side of the steps. Philip "took" him, and then set himself to studying the decorations of the outside of the portico.

The pillars supporting the arched doorway had in relief upon them forms of aquatic life, modeled life-size or larger, and arranged in geometrical patterns. For instance, one pillar was covered with frogs arranged in diagonal lines crossing one another so as to form diamonds. Others in the same doorway showed turtles, snakes and lily-leaves, newts and crabs.

Philip also saw that all of the ornamental work about the building was composed in the same way. He thought it amusing in a temporary building, but felt sure his friend the architect whom they met on the Whaleback would never have approved of the decorations if applied in equal profusion to a permanent building.

Inside the aquarium wing he found a circular corridor both side walls of which were made up of tanks filled either with salt or fresh water. To keep this water fresh and wholesome for the fish, spurts of water shot down through the surface from above, making a silvery fountain upside down at the top of each tank. There were no windows in the corridor, all the light coming through the water from the top of the tanks. This arrangement made that part of the building rather gloomy, but enabled the people to see the fish under the best possible conditions.

The people seemed to enjoy the show very much, and had none of that bored air with which they walked around some exhibits in other buildings. Each aquarium was like a show-case, and the light playing upon the moving fish caused them to glitter and shine. Philip heard one girl exclaim as she entered, "Um—um! How lovely!" and wondered for the hundredth time at the queer adjectives girls apply to what pleases them.

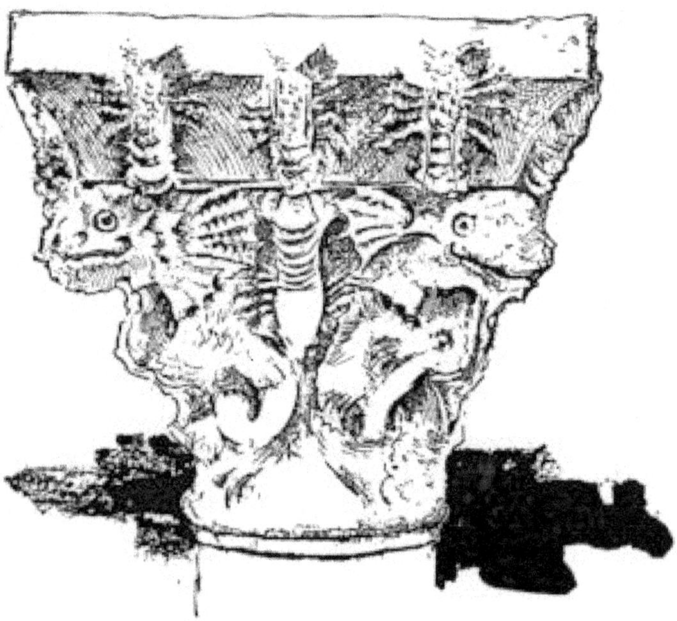

CAPITAL IN FISHERIES BUILDING.

The building was jammed full. Judging from photographs he had seen, Philip was sure that in the earlier days of the Exposition there was a better opportunity for examining exhibits. Now, everywhere he went, there were such throngs of people that he found it difficult to use his note-book. Every time he entered a building, he found his camera a burden and a trouble; but no sooner was he out again than he was glad he had brought it with him.

Here he had to fall into line if he wished to make any progress at all. People would gaze upon some slab-sided, pop-eyed fish until they entirely forgot they were keeping others away. Then the crowd would move forward with the gentle force of a glacier, and progress would begin again.

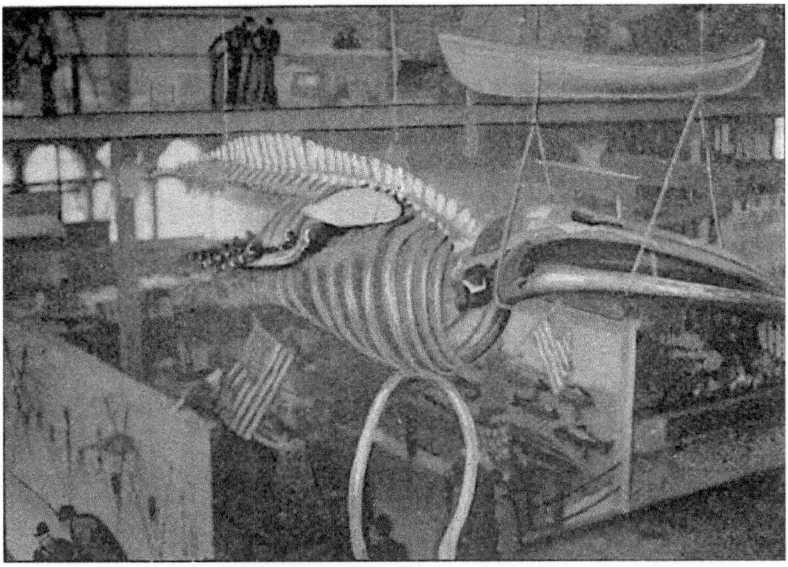

SKELETON OF A WHALE.

The first tanks Philip saw held various sorts of bass, sturgeon, trout, and pike. It was a strange sensation to see the fish so near, and so confined that they could not dart away. It gave one the idea that swimming in the sea was not so very different from flying in the air, except that a bird has to keep moving or descend, whereas the fish can stop where he pleases, and hang suspended as comfortably as Mahomet in his coffin, or more so.

FLYING-FISH.

Other fish he saw were the sheepshead, who had the true sheepish expression; catfish, with their odd sparse whiskers; some strangers labeled "small-mouthed buffaloes" (Philip wondered how it would sound to go into a restaurant and order "a fried small-mouthed buffalo, please, for one"); something that was written down in his note-book as "red-horse"—but what the creatures were like, and what their true names might be, Philip had no recollection at all when he read over his notes. There were, though, some whose names did recall exactly their appearance,—the "short-nosed gars," for instance, who had particularly long noses.

The tank of goldfish was really "lovely," for as one approached them the light shone upon them as brilliantly as if they had been sunset clouds. One visitor was so impressed with this fine display that he remarked with more feeling than logic, "It is wonderful what human flesh can do when they put their heads together!" Philip laughed at this, and after having had his elbow joggled four or five times, succeeded in writing it down so that he hoped he could tell Harry about it.

A FISHING-BOAT: GROUP IN GOVERNMENT BUILDING.
The nearer fisherman has woolen rings upon his hands to protect them from the line.

Harry and Mr. Douglass were at the other end of the grounds; for Harry, in view of his father's letter, felt that he ought to go through the little convent, and Mr. Douglass found that end of the Fair full of interest.

It was oppressively warm in the Fisheries Building, and Philip, often over-prudent, had carried his overcoat with him. He had heard so much before coming to Chicago of the "cold breezes from the lake," that he hardly dared to enter the grounds without some protection. At first the coat was light enough, but as time went on it seemed that each moment dropped a leaden bullet into one of the pockets, and his arms ached though he changed the burden continually from one to the other. His camera he made use of as a desk, fastening his note-book to the top of it by putting a rubber band around one end, but the coat became a great nuisance before the tour was finished.

In one tank he saw a queer turtle whose flippers were so broad that they reminded Philip of four fans on the ends of a frog's "arms" and legs. The sand-pike, the golden ide ("A queer way to spell it!" said an old lady), and the Missouri catfish occupied more tanks, and claimed Philip's attention in their order. The last tank of the outside row was filled with minnows,—such as boys call "shiners,"—and reminded him of a big bait-box.

Then came salt-water fish, and the change in the color and clearness of the water was very noticeable. It was much harder to see the fish, and when they scooted off to the other side of the tank, they were lost to view. The lobsters had a sort of pile of rocks to which they clung fast, and the crabs also seemed disinclined to move about. When he came to a tank labeled "sea-robins," it was some time before Philip could see why the queer little fish were so called. At last, when one came near, he noticed a red spot beside its head, and concluded that this accounted for the name. One of the oddest specimens he saw was called the "paddle-fish." It had a long flat nose extending out flat forward—probably it was used by the fish as a sort of shovel to stir up the sandy or muddy ground where its food was found.

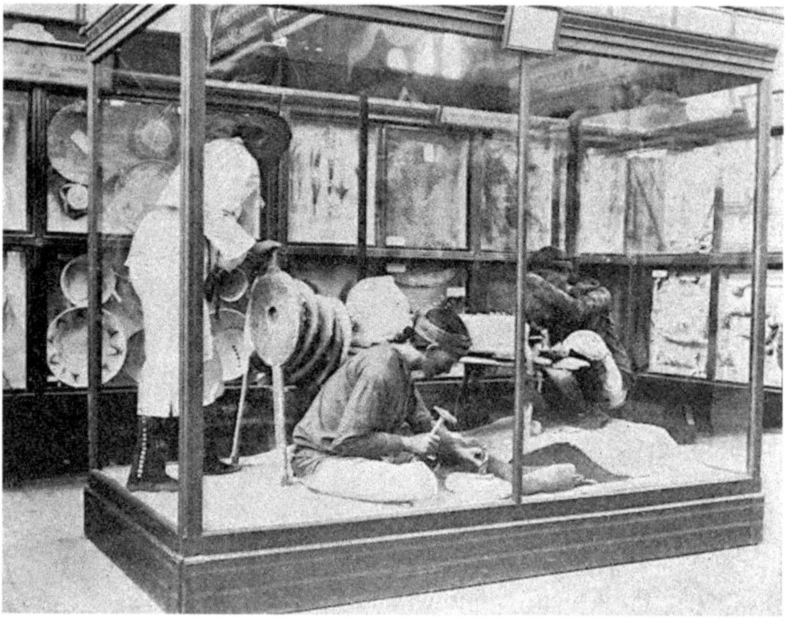

MODEL OF A GROUP OF INDIAN METAL-WORKERS, IN THE GOVERNMENT BUILDING.

Philip wondered what the fish thought of their queer situation. Instead of having the whole ocean to roam through, they found themselves in narrow quarters around which great animals with staring eyes pressed continually. They did not seem at all frightened, and had probably given up their situation as a problem the solution of which was not meant for them to know. At least they must have found some satisfaction in the absence of the enemies who usually chased them about without regard for Sundays or holidays. Philip, who was of a speculative turn of mind, wondered how it would seem to men if lions and tigers might at any moment come around the corner to devour them. He hoped that the fish were less sensitive, or he was sure their lives in the ocean would be so unbearable that they would commit suicide by leaping out on shore.

MODEL OF AN INDIAN WARRIOR: GOVERNMENT BUILDING.

"Them's catfish."

"Oh, no, they ain't."

"I tell you, they be."

"Wal, I guess not."

"Wal, I guess yes."

"Oh, you go 'way; I guess I know!"

The scientific discussion given above had proceeded no further when a cooler-headed member of the party pointed with a peace-making finger to the label, which read "Catfish," as plain as print. Hoping that these visitors' knowledge of fishes had been improved by this little difference of opinion, Philip found that he had exhausted the contents of the outer corridor, and went into the middle, where he found a rockwork fountain surrounded by a pool full of other fish. He went around the tanks seen from the middle of the building with the same care he had given to those outside, and found plenty to pay him for the trouble.

MODEL OF A GROUP OF ZUÑIS: GOVERNMENT BUILDING.

In one compartment were several sharks, and affixed to one of the sharks were two of those fishes called "remoras," who have upon their heads a sort of sucker that can be used to hold them to any smooth surface. Philip remembered reading that the ancients thought these fish could stop even a large galley. He had always regarded the statement as a wild yarn of antique romancers, but he was glad to see just how the remora applied himself to his vocation. The shark was unable to get at his unwelcome guests, and there were two of them, each more than half as long as his host. Philip said to himself that it was a shame, and then he happened to think that it was not necessary to be very sorry for sharks—which are not a kindly race. What the remora had to gain by this attachment he couldn't exactly see, unless it was mere transportation from place to place. Possibly the shark would leave something of every meal, and then the remoras would dine at the second table. It was as if a banker should have two professional beggars sit upon his shoulders, and pick up the odd change that he didn't look sharply after.

ARMY WAGONS, WAR DEPARTMENT, GOVERNMENT BUILDING.

The next remarkable fish that attracted his eye—or rather, repelled it—was the file-fish. This creature, if it *was* the file-fish, had a strong family resemblance to an unequally cooked and lumpy buckwheat cake, and was hardly thicker. It was an animated pancake swimming edge up. But what interested Philip was its method of propulsion. Along its back ran a fin for nearly the fish's whole length, and this fin waved in a curving line like the path of a serpent. Philip had heard Harry wonder why ships were not propelled by some such device, and he resolved to tell his cousin that Nature was ahead of him in using that means of going through the water.

Then Philip walked along the curving corridor with ornamented columns that led to the main building. Just as he entered this part of the central hall, he saw a clever bit of advertising. It was headed, "They say it's hot in Southern California," and below was a statement of the daily temperature contrasted with that of Chicago. For that day the California temperature was 67° as contrasted with Chicago's 73°.

GUNS, TORPEDOES, AND FLAGS: GOVERNMENT BUILDING.

Philip did not find this main building as interesting as the aquarium part. There were many models of fish, but they seemed very tame after the live ones. In the Netherlands exhibit (as, indeed, in most of them) was a model fishing-boat, but Philip did not know enough entirely to comprehend the purpose of the different devices shown, so he gave them only a glance. The exhibits of nets were likewise of small interest to him, though a fisherman would, no doubt, have been long entangled in their meshes.

The red disk on a white field that again marked the Japanese show promised him more entertainment, and he entered the inclosure. Here he found several fine little models, the most novel being that displaying the method of fishing with cormorants. A little boat full of fishermen was upon the painted waves, and in the bow was a torch made of an iron basket wherein flamed some material that had been soaked in oil. In the model this was represented by dyed wool. Each fisherman held in his hand a cord fastened to a ring fixed tightly around a bird's neck. The birds were swimming about and diving for fish. When a fish was caught, the bird was hauled in, deprived of his prey, and sent out to try again.

THE WORLD'S FAIR POST-OFFICE: GOVERNMENT BUILDING.

AN OLD-FASHIONED MAIL-COACH: GOVERNMENT BUILDING.

"FURTHEST NORTH": GOVERNMENT BUILDING.

There were in cases different kinds of fish-hooks, twisted and turned into all European shapes, besides some eccentric ones of their own, spoons and other devices for trolling, snells and lines, not very different from those used in America and Europe. Their sail-boats differed, however, from ours in the way the sails were made. Instead of being in one piece, the sails were in perpendicular strips fastened together by a network of cords so as to leave open spaces.

Philip saw a young Japanese (he looked young, but may have been fifty) who was eating lunch in a corner of the room, and asked him the reason of this arrangement. "To hold wind less," he said; but the American boy was not quite satisfied, for he could not see why a smaller sail would not meet the same need. He thought it more likely that the sails were so made in order to stow away more easily. The Japanese boy saw nothing queer in the boats, but Philip's camera was to him a great wonder, and he politely asked an explanation of its working. This Philip gave, and took the little Jap's picture in the course of his lecture on cameras. He also gave the foreigner a memorandum of the name and price of the camera, whereupon courteous Japan presented a catalogue of the exhibit and a business card.

THE BIG TREE: GOVERNMENT BUILDING.

In the main hall the State of Washington had hung an enormous "humpbacked whale" skeleton nearly forty-eight feet long, and showed the jaws of another as a gateway to its inclosure. Norway showed great harpoons and guns to project them. Baltimore, Ireland (a critical passer-by said, "How very Irish to have a Baltimore in Ireland!"), showed a model fishing-school, a set of tiny buildings with little dolls at work making nets. The dolls' idiotic faces took away all likeness of the exhibit to nature; and Philip, just from the tiny Japanese fishermen, so perfectly modeled, thought the difference spoke strongly in proof of the artistic sense of Japan.

Philip examined the models of German fishing-craft, and was particularly curious to know about a small boat moored to a tiny tree, one of three trunks below the surface of the water. He consulted the label, and found out that this was a "*Miesmuschelzucht in der Kieler Bucht*," and with that information written down carefully he departed, satisfied to wait until he had more time and a German dictionary.

ORDNANCE DEPARTMENT, UNITED STATES ARMY.

More netting exhibits—"strings and things"—did not long delay Philip, who had caught sight of the space covered with green cloth where Gloucester, Massachusetts, had arranged her boats and buildings so that one could understand how they contributed to the comfort of mankind and themselves. A lobster-packing house had made the same attempt to inform the world just how the poor lobster came to be caught, canned, and sent to table; but here some cheap dolls again marred the effect of the well-made apparatus.

North Carolina showed a "rush camp," a round hut of rushes in which had been put the proper fittings to show what accommodations their fishermen made for themselves. Mexico had a display that may have been worth seeing, but Philip noticed the fence only, which was a clever bit of work. As he left the Fisheries Building, he felt that, like the others, its display was too good and too full to be appreciated by any but experts—for whom, probably, it was especially prepared.

MAIL-SLEDGE AND DOGS: GOVERNMENT BUILDING.

He felt sure that every man or boy who went to the Fair saw some device or method that he would either adopt or improve in his own work. With a people so quick of apprehension and so inventive as Americans, the benefits arising from the World's Fair must be beyond exaggeration.

After leaving the Fisheries, Philip made up his mind to give the Government Building a good two hours of his day. He had passed through it several times, but he had never examined thoroughly the guns and wax Indians and mail-wagons which seemed especially provided for the delight of boys. Now he was glad that he had saved up the pleasure.

The Government Building was as crowded as the Fisheries had been, but Philip pressed slowly along, catching sight first of a fishing-boat and the figures of two men in it arranging their shad-nets. The Patent-office exhibit, which he had promised himself much joy in looking over, he found almost too confusing, as had Harry before him. So he passed quickly through this section and reached the exhibit of the Post-office Department, where one could see at a glance every possible way of carrying the mail, from an old stage-coach to the latest mail-car.

The Smithsonian Institution and the Ordnance Department of the United States Army exhibited what Philip felt were really just the most interesting things he had seen in the whole Fair. The groups of wax Indians, the great guns, the army-wagons, and the dog-sledges were surrounded by groups of delighted people of all ages.

Then Philip decided that he would go to the Japanese tea-house, taking in the beautiful model Japanese house on the Wooded Island. He found the model house, but it took him fully twenty minutes to find the tea-house, with four consultations of his map; and while seeking it he saw the Brazilian Building for the first time, although he must have passed it again and again. This will give some idea of the size of the Fair, for that building is 140 feet high, 148 feet long, and of equal width.

In the Japanese tea-house Philip sat upon a wicker stool, and received a cup of "ceremonial tea," a half-pound of the tea, a wafer, some sweetmeats, a souvenir, and elaborate courtesy. He also received a ticket entitling him to enter the tea-houses where the cheaper tea was served. After a long rest within this pretty inclosure, Philip took the electric launch to the southern part of the park, where he wandered about, taking an occasional snap-shot, until he felt his legs would no longer submit to be imposed upon. He went home very weary; but he was getting used to that.

THE JAPANESE "HOUSE OF THE PHŒNIX" ON THE WOODED ISLAND.

PORTRAIT OF COLUMBUS, BY LORENZO LOTTO, 1512.
In the Convent of La Rábida. Lent by James W. Ellsworth.

CHAPTER XIV

The Convent of la Rábida — Old Books and Charts — Paintings — A Fortunate Glimpse of the "Santa Maria" — Portraits of Columbus — The Cliff-Dwellers — Cheap Souvenirs — World's Fairs in General.

AN ANCIENT
CARAVEL.

AS has been said, Harry and Mr. Douglass set out for the Convent of Santa Maria de la Rábida, which means "Blessed Mary of the Frontier," according to the wise men who write guide-books. Appropriately built upon a point of land, it was surrounded by green turf to the shore, where pointed rocks made an irregular wall. Even to one coming through the Court of Honor—an architectural display unequaled—the quiet little convent presented an aspect of quaint simplicity that was full of dignified repose. Its plain walls and low-pitched roof were relieved only by two features that broke the sky-line, a tower and a belfry. Probably its designers thought little or nothing of architectural beauty, and had attained their object when they had made an inclosed court surrounded by small rooms, with one or two large enough for a refectory and chapel.

Entering a narrow doorway at the back, Mr. Douglass and Harry found themselves in a large hall, which was no doubt the chapel of the original building. To their right was the place where the altar had stood, but in the model this inclosure contained pictures on the walls. They were very old, no doubt; but when a Columbian guard told an inquirer that they were "more than a thousand years, I guess," Mr. Douglass and Harry concluded that the guard's uniform was no guaranty of his knowledge.

THE ORIGINAL CONVENT OF LA RÁBIDA, IN SPAIN.

In the front of this chancel was an easel sustaining a frame that protected the commission authorizing Christopher Columbus to go and see what he could find. A placard requested "gentlemen to remove their hats," as Philip had said; but the American public had made up its mind to disregard this inscription. Mr. Douglass said to Harry, in a low tone, "I can see no reason for removing one's hat to a piece of paper with ink on it. One can show a proper respect and appreciation for a relic without flunkeyism." And Harry quite agreed with him.

The commission was a bit of brown parchment written in a crabbed hand, probably by some court copyist; and not even the signatures were intelligible.

Moving onward through the crush of people, they came next to the west wall, where there was a glass case containing the rarest ancient treatises upon geographical matters. There were twenty or thirty in the case, some ornamented with woodcuts; but though Harry had come with the best of intentions to study the exhibits carefully, he could do nothing but gaze wonderingly at the type, saying to himself, "This is an old, old book. Columbus may have read it. Here's another. What a queer picture!" At length he said to Mr. Douglass:

"What do you make of them, Mr. Douglass?"

"Very little, I must confess," said the tutor. "One has to read such books to learn how much wheat there is amid the chaff of fable, folly, and guesswork. Even if I could read all the languages, I could get little from the two pages which are all they can show. All you can do now, Harry, is to get a good idea of what these old books and charts are like. Perhaps we can buy a catalogue which will give us translations of some parts of the books and of the letters that are also shown here."

"What can one learn from these old books?" asked Harry. "Surely there is nothing in them that we don't know about."

THE CONVENT OF LA RÁBIDA AT THE FAIR.

"No, of course there isn't," said the tutor. "We can only take an interest in them as showing the beginning of events that have resulted so wonderfully. It is also true that now and then we find a sentence throwing light upon how men did things in old times. But it is rather as a matter of curiosity than of learning that these relics are studied."

Upon the west wall was a very large painting showing the "Landing." It represented Columbus, just after he had stepped ashore, raising his eyes upward in thanksgiving for his success. The men in the boat seemed to show curiosity and enjoyment rather than piety. The painting was not remarkable except that the expression upon the discoverer's face was well rendered.

At the left of the picture the original anchor of the wrecked "Santa Maria" leaned against the wall. Both flukes were gone. Mr. Douglass felt a little doubtful of the genuineness of this relic, but was willing to be convinced. There seemed to be no proof that the anchor belonged to the old caravel; but, on the other hand, it was found where her anchor might have been, and it was pleasant to believe that it might be the very piece of iron upon which the hand of the discoverer had often rested.

CELL OF THE PRIOR MARCHENA IN THE ORIGINAL CONVENT,—THE "COLUMBUS ROOM" IN THE MODEL AT THE FAIR.

Mr. Douglass asked a young man who was selling catalogues whether he hadn't one telling just what letters and papers were in the building; but, to the tutor's disappointment, the catalogues of the old manuscripts were all sold. There were a number of old paintings around the walls, but after examining a few, Mr. Douglass advised Harry not to waste much time over them, as their connection with Columbus was rather remote.

Going up the narrow stairway, they came out upon a corridor that looked upon the open court through graceful arches; and from this corridor opened small square rooms—originally the monks' cells. The doors were as plain as possible, and each had a latch lifted by a string coming out through a hole in the door; the windows were small and square, embrasured, or sunk, into the thick walls.

HOUSE IN GENOA SAID
TO BE THE BIRTHPLACE
OF COLUMBUS.

Various relics were hung at every point, either along the corridor or in the cells. Many of them were queer drawings or paintings, meant to show the manners and customs of the Indians; others were charts and maps, some earlier than Columbus and others later.

At one place Harry suddenly stopped and pointed delightedly out of the little window.

"See!" he cried. "Isn't that just right?"

Mr. Douglass turned, and gazing through the little opening saw the "Santa Maria" lying near the wharf opposite.

"It makes one feel as if he were back four hundred years," said the tutor, quite as much delighted with this fortunate view as Harry was. "Looking through this little window, we see nothing of the crowd, and are all alone with the convent and the caravel."

They were most interested in the "Columbus Room," which occupied the place of the cell where lodged the monk who became interested in the man with a theory. There were dozens of portraits of Columbus, and they certainly gave one plenty of choice. Broad-faced, narrow-faced, round, oval, bearded, or smooth, the great discoverer might well have been puzzled to know which was his likeness. People's remarks were droll enough.

One young woman who had been critically scrutinizing the array of "Columbuses, various," finally stopped delightedly before a large portrait and exclaimed:

"Oh!—*that's* more like him!"

Harry longed to ask how she knew that, but concluded it would get him into trouble. Harry himself had no choice. He felt just as another critical visitor did. This was a young man in a broad felt hat, who sailed around the room, and left with the parting remark:

"There isn't one of them that looks alike!"

Mr. Douglass and Harry spent a long morning in the convent, but Harry wearied of it. He tried to be interested, for he wished to please his father; but he couldn't find anything to take hold of in making a beginning. Still, by sitting quietly in the rooms and corridors, the boy, without realizing it, carried away a perfectly clear idea of the old convent, its arrangement, how Columbus must have been lodged and entertained, what the old documents were like, and how much modern maps differed from the rude charts of the Middle Ages.

He told Mr. Douglass that he couldn't make much of it; but the tutor consoled him.

"You never know how much you have learned until long after you've studied and gone past a subject," said Mr. Douglass. "Some day you'll read more about this old building and its documents, and then you'll find a peg to hang the knowledge upon. Have you ever seen a negro minstrel try to hang his hat on a wall where there is no hook?"

Harry laughed, and said he had.

"That is what people must do who have no general ideas to hang particular bits of information upon. Now, in this case you would be surprised to see how much you know about Columbus compared with what you knew before you came to this Fair. I won't bother you now to review it; but some day, when we are studying again, I'll let you note down the facts about Columbus that you learned at the World's Fair."

"Thank you," said Harry, smiling.

"You'll like to do it," said Mr. Douglass. "You'll see. Now let us take something a little simpler. I hear that the Cliff-Dwellers exhibit is really good. Suppose we go over there?"

Harry was very glad to agree, and they walked still further southward past the Anthropological Hall and the Forestry Building,—a most interesting place, where none of them had yet been,—and came to the curious imitation of a great cliff which gave room to the Cliff-Dwellers museum and models.

Here they found that there were guides to go about and explain the different parts of the show. They followed one of them for a while, but found that he talked so fast and paused so short a time in any place that they could hear and see little.

Starting out upon their own account, they looked first at models built into the sides of the imitation rock,—for they were inside a great structure dimly lighted, and looking like a great cavern,—showing that the "villages" were really a collection of rooms made by erecting walls from floor to roof of a cleft in the cliff.

DEPARTURE OF COLUMBUS ON HIS VOYAGE TO AMERICA. (IN THE CONVENT OF LA RÁBIDA.)

Then they went into the museum, and saw relics of the strange people of whom little is known. Some believe there are no remnants of these Indian tribes of the Southwest. Others think the Pueblos are the same or a closely connected people.

A LAMP.

There were in the cases bits of sandals woven of cord, cloth remnants, some as finely woven as canvas; bits of bones, scooped out into spoons or sharpened and faced for needles; bits of straw hats, large stone mortars for grinding corn, the corn itself in jars and corn-cobs, and even skeletons, skulls, and mummies in a fair state of preservation. The skulls were finely developed in front, but nearly all flattened at the back. The skull of a Cree Indian was set in the case, in order to show how much finer were the foreheads of the Cliff-Dwellers. Harry was especially pleased to find a little bear made out of pottery,—a tiny little thing that was probably a toy. He made a sketch of it.

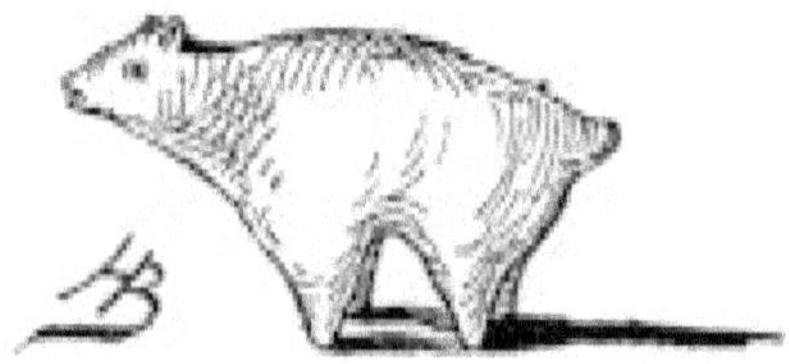

Going into another part of the cavern-like structure, they saw some oil-paintings of the original rocks and dwellings from which these relics were taken. In yet another compartment were some of the donkeys used by the exploring party, and young girls and children took rides upon their backs.

Harry was standing just at the heels of one of these little gray beasts of burden when a gentleman of an inquiring turn of mind asked, "Does he kick?" at the same time pinching the donkey to see. Luckily for Harry, the donkey didn't kick, or there might have been a new mummy added to the collection in the museum.

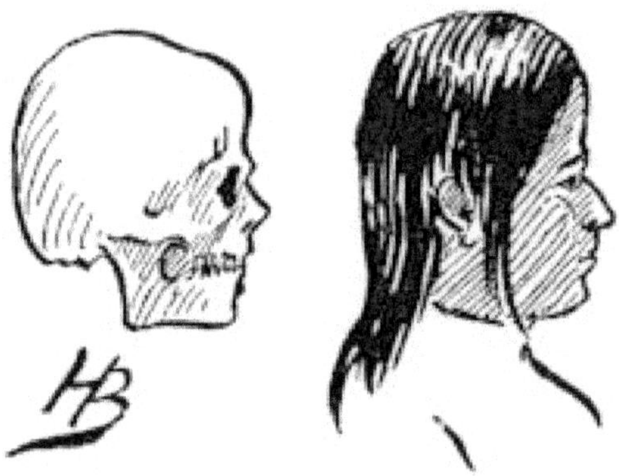

HARRY'S RESTORATION
OF A CLIFF-DWELLER.

Another place was set apart for full-size models of the houses, and it was curious to see how the walls surrounded a sort of fireplace. The Cliff-Dwellers apparently slept in stone bunks cut in the rock, for there were several of these mineral sofas around the walls.

A path here and there led up out of the interior to the surface of the imitation cliff, and many people went clambering up and down these strange highways, clinging to ropes that had been fastened along the sides.

Now and then the lecturing guide would come near enough to let Mr. Douglass and Harry hear part of his explanation. He spoke of the deep wells that supplied the dwellers with water; of the narrow trails that made the settlements inaccessible to the hostile tribes that drove this people from the fertile plains up into these rocky forts; of the lamps and the cooking-utensils:

but all at such lightning-express speed that Harry could find out nearly as much by examining the objects for himself.

When they came out, Harry was amused to see that even the turnstile was made of rough logs, to be in keeping with its surroundings.

THE CLIFF-DWELLERS' MOUND.

As Mr. Douglass intended to go out to see a procession of boats in the evening, they went home early. On their way they passed the Alaskan totem-pole settlement, but concluded not to make their way through the press in order to get into the hut where souvenirs were on sale. At the Indian village they did succeed in making their way within doors, but found nothing to repay them for their trouble—merely the regular array of baskets, bows and arrows, and similar trifles.

Some North African booths, kept by people of the French colonies, offered for sale all sorts of little trinkets in brass and silver filigree or cheap enamel; but in spite of the continued cry, "Sheep, sheep; everyt'ing werry sheep!" the party kept on toward the outlet.

Reaching home, they found that Philip was already there. He was at the window, much amused over the doings of some of the negro waiters who, sitting around in the sunshine, were musically singing or talking over their experiences.

Philip displayed his photographs, and Harry showed the sketches he had made. But the party had ceased to be very talkative over the Fair.

"What I should like to see," said Mr. Douglass, "is a fair from which all the ordinary, commonplace exhibits are excluded. Cans and boxes of ordinary merchandise, even if piled up in ornamental forms, are better suited to an agricultural county fair than to a World's Exposition. A small, choice exposition, where every exhibit was unique of its kind, would be more manageable and much better worth seeing. This Chicago World's Fair has in it the very best material the world can produce. But it would take two years to see it thoroughly, and no one man could understand it then."

"I'll tell you what I should like to see," said Harry; "and that is a grand procession where people of the same States should be in ranks together. Then we should see how they differed."

"And my idea," added Philip, "is to have a Children's Fair, where everything that is interesting to boys and girls should be on exhibition. That would be something like!"

VIEW LOOKING SOUTH FROM THE TOP OF THE WOMAN'S BUILDING—BY MOONLIGHT.

CHAPTER XV

The Electricity Building — Small Beginnings — A New Souvenir — The Curious Exhibits — Telephones and Colored Lights — The Telautograph — Telegraphy — Mines and Mining — A Puzzled Guard.

"IT is interesting to reflect that the beginnings of all the marvels we shall see in this building," said Mr. Douglass, as he walked with the two boys toward the Electricity Building, "are found in two trifling circumstances that the majority of men would have overlooked. Do you remember what led to electrical research?"

"I know," said Philip, "that the word comes from the Greek for amber, and I suppose you mean the attraction of amber for little things was one of the two."

"Yes," said Mr. Douglass. "Now what was the other?"

"Frogs' legs," Harry answered. "I remember reading about that not long ago. Volta salted the frogs' legs, thinking they were too fresh; and they kicked. That's what you mean, isn't it?"

"Exactly," said Mr. Douglass, laughing. "And that frog-kick was the beginning of the impulse that laid the Atlantic cable. It was no doubt a great achievement to come upon a new world, as Columbus did; but really Volta, who knew exactly what he was about, deserves nearly as much credit. So you see that by carefully noticing what takes place in his own home in the course of his every-day life, a man may become renowned quite as well as if he braves the elements in search of a new continent."

"Do you think electricity will take the place of steam?" asked Philip.

"No," answered Mr. Douglass; "for, judging by the past, few really useful things are ever displaced. Every housekeeper still finds a need for candles, even where not only gas but electricity is at hand. The stage-coach is still built and used, though for different purposes than at first. We shall see to-day, in the Transportation Building, how many old inventions are yet on duty."

THE ELECTRICITY BUILDING.

As they entered they heard a sharp pounding, and saw a crowd gathered—the surest sign of something interesting—near a counter. Gradually making their way to the front, they saw a sign announcing that they could have their own coins made into Fair souvenirs, and found upon the counter small scarf-pins, medals, monograms, hair-pins, and paper-knives made from silver and nickel coins. The charge was only five cents, so Philip drew forth a half-dollar that he had been intending to spend on a present for his sister, and putting five cents with it, handed the coins over to the woman at the counter.

"What would you like?" she asked.

"A hair-pin like that," said Philip, pointing to one that had 1893 upon the top in openwork. The woman gave the half-dollar to the man at the stamping-machine, and he pushed it under the die. In a few moments Philip's coin was transformed beyond recognition, and came out properly shaped and labeled "Columbian Exposition, 1893." Harry satisfied himself with a nickel rolled into an oval and also stamped.

PORCH OF ELECTRICITY BUILDING.

A little further on they saw a counter where handkerchiefs were embroidered with appropriate inscriptions, also to serve as "souvenirs"—a word of which the party were becoming weary, as it was bawled, shouted, and whispered in their ears from morning until night.

Many of the electrical exhibits were interesting only for their arrangement: there were, for instance, carbons arranged in geometrical patterns, and push-buttons forming letters and inscriptions.

It was not until they had reached the southern end of the building that they began to think well of the electrical exhibition. But toward this end the attractions were most striking. There was a whirling ball of electric lights, hung near the ceiling, that Harry remembered noticing on the first evening, when they had so much trouble to get in and out of this building. Not far from this ball was a column of colored-glass lamps, from the top of which lines of lamps ran zigzag over the ceiling, each ending in a hanging lantern.

This column would suddenly gleam with colored fire at the base, then further up, then to the top, the waves of light dying out below as they ascended. Reaching the top of the column, the zigzag lines flashed out in wavy lightning flashes to the hanging lanterns. Then all would become dull, until another impulse made its tour of the line.

Another beautiful exhibit was an Egyptian temple. The pillars were of roughened green glass lighted from within so as to glow like emeralds. The walls contained show-cases displaying electric fixtures.

The boys had heard praises of the electrical theater situated in this corner of the building, and it was one of the places they had made up their minds to visit. But they found a line of people ranged before it, and extending back far enough to discourage any but an electrical crank. Reluctantly they withdrew, and went instead into the Greek temple, where a telephone was in working order. A row of young girls sat upon high stools facing a bewildering array of pegs. Upon their heads the girls each wore a light frame of metal bands that held telephones to their ears. It was a striking illustration of the line about "lend me your ears"; but in these modern days the ears are hired by the week. Every now and then one of the girls would lean forward and change a peg from one place to another.

Besides the receiving instruments, a transmitter hung down just in front of the lips of each operator. In fact, every care was taken to enable these young women to hear all conversation addressed to them, and every facility given them to answer back.

Harry said he thought it was just the sort of work a girl liked—nothing to do but to be talked to all day, with full liberty to talk back from a safe distance; but Mr. Douglass said that he had heard the work was very hard and exhausting.

In the gallery they found a number of amusing or astonishing novelties. One that Philip found attractive was an electric boot-blacking machine. In front of chairs like those belonging to the regular "Have-a-shiners" of commerce, there were two brushes revolving rapidly. A man sat in the chair applying his well-developed foot to the brush, and receiving an electric shine that was nearly as good as the regular article.

Harry watched this device critically, and at length said he didn't like it.

"Well, I do, then," Philip answered. "Wouldn't I like one to use every morning, though?"

STATUE OF BENJAMIN FRANKLIN AT THE MAIN ENTRANCE OF THE ELECTRICITY BUILDING.

“I mean that the principle isn’t right,” Harry insisted. “That inventor is making the man twist around so as to apply his foot to the brush. He ought to make an electric brush that can be held in the hand and put against the boot. Don’t you think so, Mr. Douglass?”

“Your argument seems reasonable,” said the tutor; “but it’s often wise to remember that the inventors have thought more about these problems than we have; so it is not likely they have overlooked the most evident criticisms. Still, in this case I think Harry is right.”

MODEL OF A LAKE SUPERIOR COPPER-MINE: MINING BUILDING.

At another place in the gallery there was an electric door, and people were invited by placards to walk through it. It had a handle like other doors, but no one ever touched it; for no sooner did one approach than the door opened politely, closing after the person was upon the other side.

One man—"who thought he was smart," Philip said—walked up to the door as if he meant to pass through the doorway, and then halted. The door remained open so long as the man stood before it, and closed when he turned away.

"It seems a pity to fool a door that is so polite," Harry said. "Look," he added; "there is a nice little girl trying it. See her laugh! It reminds her of 'Alice in Wonderland.'"

MINES AND MINING BUILDING.

Germany had a historical exhibit showing the earlier and cruder forms of dynamos; but the boys were not very well acquainted with dynamos. Mr. Douglass tried to explain how they worked; but after he found he had lost the trail of his ideas, he said frankly: "Well, I thought I knew the theory of dynamos and converters; but when I see the real machines here, they seem so much more complicated than the ones in the text-books that I find I don't know the reason for many of the parts."

AN EXHIBIT OF RAILS: MINING BUILDING.

The boys took more interest in the Western Union exhibit, where they saw Professor Morse's earliest receiving instrument, and photographs of the original first message, "What hath God wrought!" The same words were affixed to the front of the pavilion, where not only the original instrument but the modern quadruplex system—a method of sending two messages each way, and all at once, on a single wire—was shown.

"I wish," said Harry, "that I could see the game of leap-frog these quadruped signals must play to get by on the same track!"

Farther on were other German or Austrian exhibits, in one of which the boys saw a dome copied from that on some central telegraph station, and made up entirely of openwork so as to give room for hundreds of insulators. These insulators made up the curved surface of the dome, and the effect was very decorative, while the arrangement must have been a great saving of space.

What a lot of things there were besides! There was an electric cooking-apparatus where water was boiled upon a flat iron plate; there were clocks so contrived as to note the times a watchman touched a button on the front; there was Professor Gray's telautograph, which merits some description.

TWISTED IRON: MINING BUILDING.

Holding a pen as in writing, the sender marks down his message, draws a design, inscribes his name—in fact, uses the pen as freely as if it had "no connection with the establishment across the way." But two cords extend out from this pen and work an electric apparatus so as to pull two other cords or wires just as the first ones are moved: if he makes a mark down, the other pen is pulled down too; whatever one pen does, the other must do. Of course, then, any drawing or writing made upon one machine is also made on the other—no matter whether it is in the next room, the next county, or the next State. That is the telautograph—the name being Greek for "far-self-writer."

In the exhibit of the Commercial Cable Company were shown the method of writing messages in wavy lines, and bits of cable where the covering had been injured, and the injury—sometimes no larger than a tack would make—traced and located many miles from shore by means of delicate tests.

Down-stairs were great dynamos, electric cars, the Edison-light tower, which they had already seen in operation on their first evening at the Fair, and such an array of complicated measures, meters, and tests that the boys walked humbly out, feeling very small indeed as they passed the heroic statue of Benjamin Franklin in the portico. They felt that for the first time they understood how great a man was the printer's boy who began by carrying two rolls under his arms and ended by carrying a thunderbolt under one arm and a scepter under the other.

"But even he," said Harry, as he jingled a pocketful of expensive souvenirs, "once paid too dear for his whistle."

The Electricity Building's stocky twin, the Mining Exhibit, was right next door, and came next upon Philip's neat list. But they did not intend to give a very long time to this building. They knew it to be full of minerals and mining machinery, and now felt small enough to admit there were two or three things in each display that they did not understand.

The first distinct feature was the Stumm exhibit, which, behind a most imposing gateway of wrought-iron, showed rails and pipes in sizes ranging from mammoth to midget, built into two towering obelisks, and two trophies that resembled iron fountains. They gazed upon these with vague admiration, and then set out to find the Tiffany diamond show; they "found it, indeed, but it made their hearts bleed" to see the crowd piled three or four deep against every loophole and knot-hole where a wheel or a band was visible.

The same result followed an attempt to inspect the Kimberley diamond-washing. They did see an enormous Zulu with embroidered suspenders pour a bucket of bluish mud into a great hopper, but though they lingered round in a most lamblike way, nothing else was to be observed.

Iowa showed a life-size model of a coal-miner at work in his gallery; and at one glance the boys learned how it would feel to be "down in a coal-mine, underneath the ground, where a ray of sunlight never can be found." They also enjoyed hearing and seeing the steam-drills, and gazed curiously at a model of "Lot's wife,"—a woman built of salt,—in the Louisiana Exhibit. Various mines had sent models showing just how their galleries were built, and the boys inspected them critically. But they did not find very much to detain them in the Mining Building. Other people, too, seemed more interested in the souvenir stands than in the profusion of ores and stone blocks. Montana's silver statue of Justice seemed to the boys more of a curiosity than a work of art, and they had no patience with the long arrays of machinery that meant nothing to them. Those who were examining the exhibits were few, and the large crowds were watching the counters where small metal articles were plated, or were sitting in corners where they could rest themselves.

A Columbian guard noticed that Philip had his kodak, and said, "You can't take pictures in here; it's not allowed."

"I haven't taken any," said Philip; and then, as the guard seemed good-natured, he added, "I don't see anything much to take. Why don't they let you take things in here?"

The guard grinned. "I'm sure I don't know," he said. "There doesn't seem to be any sense in it."

SOUTH PORCH OF MINES BUILDING.

THE "GOLDEN DOORWAY" AND PART OF THE TRANSPORTATION BUILDING—ON A QUIET AFTERNOON.

CHAPTER XVI

The "Golden Doorway" — Transportation Building — An Endless Array — Bicycles, Boats, and Bullock-wagons — The Annex — The Railroad Exhibits.

THE CROWD COMING
IN WITH LUNCHES.

FROM the steps of the Mining Building the boys looked over toward the "Golden Doorway" of the Transportation Building, and made up their minds that it looked promising. By this time the white buildings had made them glad of the fancy harlequin costume worn by the autumnal-colored member of this interesting family. They liked even the angels painted along the walls, and as for the brakeman, "Mr. Land," they thought he appeared to be a young fellow well worth knowing.

So they entered with a readiness to enjoy whatever they should find. But they soon discovered there was no need to make excuses for the Transportation Building, and before long they carried out to the letter Harry's punning prediction, "Now we shall go into transports!"

They had missed so much at other times by leaving the galleries to the last that this time they went at once up the stairs. But on the landing they turned to take a view of the Lord Mayor's Coach, an elegant turnout, as fine as a fiddle, which made the boys think at once of poor little Dick Whittington.

It was Harry's proposal to go into the gallery, and he was led to make it because there were set upon the gallery-railing two bicycles, ridden by dummy figures of a young man and a young woman. Harry liked bicycles, and meant some day to have "a beauty"; and he thought this was a good opportunity to get points.

FIGURE OF BRAKEMAN,
TRANSPORTATION BUILDING.

He got points; in fact, he picked so many points that he couldn't remember them, for there were bicycles enough in the gallery to bend all the backs in a city into the letter "C." But before examining these, the whole party were glad to give some time to Mrs. French-Sheldon's camping-outfit and traveling sedan-chair. Shortly described, it was just a basket on poles, but it was sumptuously fitted up with cushions and awnings, and most ingeniously contrived so as to be light, comfortable, and convenient.

"She's the woman who collected all those odd things we saw in the Woman's Building," said Philip.

"Yes, I remember reading about her in the papers," said Harry. "She carried a fine silk dress with her, and always put it on when she received a native ruler. She seemed to think they liked it. But I have my doubts. I believe old Sultan Alkali Ben Muddy would grin when he was climbing back on his camel, and say to his first camel-driver, 'The white woman is plucky, but I must say she puts on a lot of style!'"

Really Harry could not help a feeling of great admiration for Mrs. French-Sheldon, and he would have liked to own a tent and palanquin of his own. Passing through a corridor of photographs showing "foreign scenes in New Jersey," as they heard a jocular Irishman remark, they saw next an Indian ox-cart, heavy enough and clumsy enough to make any civilized Buck and Bright weep. Then came a tobacco-hogshead to which was attached a branched iron pole, so that the hogshead was its own wheel and cart in one.

They heard a Southern girl say to her friend, "I've seen one just like that in Richmond." But she hadn't seen the next exhibit, for it was the model of an antique chariot found near Thebes, and supposed to be a racing-sulky of such antiquity as to be labeled "the oldest vehicle known."

Harry, and indeed all three of the party, wondered at its beauty and elegant finish. It was made of some smooth-grained wood and rounded into exquisite curves. Harry made a hasty sketch of it, but had little hope that he could really draw its exquisite curves when he got home.

BIT OF ORNAMENT,
TRANSPORTATION BUILDING.

Then they went on, to be stopped by some African palanquins, fitted with carrying-poles, and, in sharp contrast to the Theban chariot, an African log-canoe so rude that it looked like the Missing Link's private yacht. In close succession came vehicles for carrying such different articles as babies, dolls, and cash in dry-goods shops; but all were quite familiar to the New York boys. They found two "bicycle-railroads" more interesting, especially the one that hung from an overhead track.

"It wouldn't be surprising," said Mr. Douglass, "if we should live to see those tracks put up over large sections of the land. For the bicycle is capable of displacing almost all passenger-carriers except in special cases. You see them here in this gallery so arranged as to be ridden by one, two, or three riders, so as to carry children with their parents, or fitted up for the use of firemen or soldiers."

At the end of a gallery they found figures showing how Mexican donkeys are loaded, men carrying chairs for transporting passengers over mountain-trails, and richly attired cavaliers mounted upon finer specimens of the same patient donkeys that carry panniers.

An exhibition of leather saddles and similar wares brought them to a counter where whips were being covered by little bobbins revolving about as dancers whirl in the german. These whips were also for sale as sou—. "I wonder," said Mr. Douglass, "that they didn't offer to sell us the Cliff-Dwellers' mummies as souvenirs. They certainly would outlast most of the cheap bric-à-brac offered for sale."

Japan showed in this building only a few models of engineering-works, and the boys did not give much time to her exhibit. They were most attracted by the smaller articles displayed on both sides of the galleries: an English sedan-chair, such as they had seen in old paintings; a springless velocipede called the "Dandy Horse," and dated 1810; the small model of an old stage-coach; a wonderfully fine model of Forth Bridge, Scotland, showing a miniature train of cars hardly thicker than a lead-pencil; a modern club canoe, side by side with barbaric outrigger canoes from the Friendly Isles (maybe).

There was also a large model showing just what style of boat the fishermen used upon the Sea of Galilee in the days of the Saviour; it was a double-ended deep boat, looking as if it was very seaworthy, but gaudily painted.

The Chamber of Commerce of the Port of Dunkerque, France, had sent to the Exposition an enormous reproduction of the town and harbors, so large that each house had its tiny model in the mimic town. The boys admired this exhibit, and concluded that the money and labor expended upon it would not be wasted; for if they had been merchants they knew that it would have been impossible for them to forget what an excellent place Dunkerque must be for trading.

Another exhibit which they equally praised was that of a French steamship company which had made a "diorama," or series of life-size views, setting forth exactly what traveling by their line would be. And instead of being satisfied with inferior work, they had selected a skilled artist to paint their pictures.

One will serve as a specimen. It was a painting that represented the last moments before sailing from Havre to New York. The spectator saw before him the long dock crowded with the passengers. Here an old mother was tearfully bidding her son good-by; here a party of jolly tourists were waving handkerchiefs to friends upon the steamer. In another spot was a lonely traveler who seemed to have no friend other than a carpet-bag. And, in short, the whole scene was vividly rendered with artistic power and with feeling. There were eight of these pictures, and the boys left none unvisited.

From a little beyond this point the boys could see the full-sized section of an ocean-steamer that reached from the floor to the roof, that is, counting the smokestack; and the boys agreed to sample that section before leaving. As yet, they found it hard to get through the galleries. Just as they had made up their minds to go down the stairs, they would come upon something that must be looked at. Such was a Netherland fishing-boat, so quaint that Philip succeeded in photographing it, even though the light was anything but favorable.

THE "GOLDEN DOORWAY," TRANSPORTATION BUILDING.

Still more fascinating were the German exhibits of men-of-war—little, fierce battle-ships with rifled cannon hardly larger than darning-needles, but every detail so finely finished that it was like watchmakers' work. In this series were shown all sorts of boats, from the swift cruiser down to the tiny torpedo-boat.

"What toys men can make when they try!" said Harry, enviously. "To think of the clumsy things that are made for children when such little beauties as these are possible! Why, there are models of boats here in this Fair that are so neat the King of the Fairies would feel timid about entering them—and I wish I owned one of them, that's all!"

But there was no time to spare for enthusiasm. Folding-boats must be seen, and a gondola,—the last so exquisite in its fittings that the ones out on the Lagoon were like it as an ash-cart is like a state carriage,—and models of boats from India, whole cases of them, in all varieties and endless numbers.

Philip walked away and sat down in a corner.

"What's the matter, Philip?" asked Mr. Douglass—"are you tired?"

"I have been tired all the time I've been in the Fair," said Philip; "but it isn't that. I am getting mad. I want to see things; I want to learn about them, and remember about them. And there is no chance. It's like trying to pick out stars in the heavens when you don't know a thing about astronomy. As soon as you look at one it disappears, and you see another."

"Well, Phil," said Harry, "you know we leave for home to-morrow afternoon. Bear up—be brave; it'll soon be over now. Come and see the ferry-boat with the side taken out so you can understand it—if you have time. Why, you haven't begun to see anything yet!"

But Mr. Douglass stopped Harry with a warning look; he saw that Philip was really getting tired out. Harry took things more easily, and was less in earnest; but Philip preferred to see things in order, and to study them by system. Excellent as is this rule for ordinary cases, a World's Exposition must be treated differently. It is possible, of course, to study only one subject in the Fair, and ignore the rest; but no one ever does so. Human nature will not permit of it.

Descending to the main floor they walked up to the model of the Bethlehem steam-hammer that made an arch across the center aisle, and after some reflections upon the statistics attached to this monster, resolutely passed whole platoons of exhibits no visitor should miss.

Mr. Douglass and Harry left Philip to rest awhile upon a settee in one of the side corridors, while they went through the section of the big Atlantic Liner. Beginning at the steerage, they worked their way upward through the office, saloon, smoking-room, and state-rooms until from the upper deck they could see Philip's disconsolate form far below.

A SECTION OF A STEAMSHIP.

To Mr. Douglass, who had never crossed the ocean in one of these palace steamers, the exhibit was wonderfully interesting; but to Harry it was less of a novelty.

Returning to where Philip sat, they decided to take lunch before going farther, and went into a small space where there was a lunch-counter, some very independent waiters, and a slap-dash way of serving that added no relish to the rather poor food. But the rest was pleasant; and after lunch they felt quite able to enter the Annex, where they found another bewildering array of locomotives, trains of cars, torpedo-boats, car-seats, rapid-fire guns, and "other things too numerous to mention," as boys say in their compositions when they can't think of anything else.

THE "DE WITT CLINTON" TRAIN.

They went through palace cars, and tourist cars, and English railway-trains, and then sought relief by examining a military wagon so made as to tip up and form a steel-clad breastwork. They could not pass this, for a dummy soldier was leveling his rifle directly over the edge, and a placard said, "Halt!" in very peremptory letters. It repaid them for stopping, for they decided that it was new to all of them, and a very ingenious invention.

Then leaving the building, they made their way toward home, but were caught and held by the great express engine, shown by the New York Central. They had often passed it, but had been reserving a more careful examination until they should have seen the exhibits in the Transportation Building. Now they walked through the whole train; but they found it much like the "Limited Express" they intended to be in next day, steaming along toward New York. The "De Witt Clinton," the first locomotive used in New York State, stood in front of "999," and looked like a dwarf kobold beside a splendidly developed giant.

THE "JOHN BULL" TRAIN.

They heard some men sneeringly say, "That was the best they could do then!" and Harry couldn't help wondering how long the world would have had to wait for "999" if such narrow-minded men were its only dependence for improvement.

Crossing the broad white road, they next went into the Pennsylvania museum of old engines and railroad appliances. Here they spent more than an hour studying the curious history of railroad invention from the beginning. There was a model of the "John Bull," and of its descendants from children to great-great-great-grandchildren. Nor was this display confined to locomotives: there were a packet-boat, such as Mr. Douglass remembered to have traveled in when he was a little shaver in short trousers and velvet jacket, the still more ancient Conestoga wagon with its boat-like body and long awning, and the old stage-coach labeled "Twenty days from Pittsburgh to Philadelphia."

INTERIOR OF A PULLMAN CAR.

Besides these models there were relics—old tools, old lanterns, and ticket-punches; and systems of signaling were also illustrated. But it is impossible to recall or put down even the leading attractions of this clever little museum.

While Mr. Douglass and Harry were looking at these cases, and at the photographs showing views along the road, Philip wandered away to the other side of the room, and found diagrams, charts, and pieces of mechanism for showing the statistics of the Pennsylvania road.

Gilded blocks as large and larger than a boy's head, showed the amounts of silver paid to employees every hour. An obelisk built from tiny stones represented the amount of ballast in this great railroad as compared with the pyramid of Cheops which was constructed on the same scale just alongside. The pyramid was nowhere in comparison. A little globe with a railroad-track going around the equator and lapping enough to tie in a bow-knot showed the length of this railroad system. Two bits of rail whereon were silver dollars laid edge to edge, were meant to show the cost of the road—a sum large enough to cover all its rails with a row of silver dollars. Another globe had models of little locomotives running around it, to show the number of miles covered by trains—enough to encircle our globe every two hours. Tiny coal-carts, drawn by clockwork up from a pretended mine, taught that two and one-half tons of coal were burned every fifteen seconds.

Altogether Philip thought the Pennsylvania had "done herself proud"—except in the models of railroad-men in uniform. No one, however deeply impressed with the rest of the exhibit, would care to ride on a road run by such men as the dummies were. Philip would not have been surprised at a strike on the whole system if the men could have seen those great paste-board gawks that stood in their clothes.

For the last few days they had been really studying the exhibits instead of wandering around with an idea of being amused. As the next day was to be their last at the Fair, Mr. Douglass made no objection to their going once more to the Plaisance, where there was more fun than instruction; and with this prospect in view, they forgave the tutor for the useful knowledge they had been so steadily acquiring.

MODEL OF THE BRITISH BATTLE-SHIP "VICTORIA."
Sunk in collision with the "Camperdown."

IN THE LAPLAND VILLAGE.

CHAPTER XVII

A Rainy Day — The Plaisance Again — The Glass-works — The German Village — The Irish Village — Farewell to the Phantom City.

A BOY FROM JOHORE.

THE boys had seen a number of unpleasant days at the Fair, but their last day was the worst. It did much to reconcile them to going away. Not only did it rain in a fine, penetrating drizzle, but the wind blew a gale, and kindly carried the dampness where it could not have gone by itself. While walking outdoors, the boys saw nothing amusing in the weather. But Mr. Douglass, in order to cash a check, had to call upon one of the gentlemen whose office was in the Electricity Building; and, waiting for him, the boys sat at a window that looked out upon the Court of Honor, and then found that the storm had its funny side.

Visitors seemed to object to walking straight, and leaned over against the wind like a fleet of fishing-smacks on a rough day. The launches going northward found their propellers only a luxury, for their awnings made excellent sails. Hats left their owners' heads, and started to see the Exposition alone. Small boys and men played short-stop at a moment's notice, and became very skilful in "dropping upon the" hat as it rolled by upon its brim.

"Hats blown off while you wait!" said Harry, laughing as he saw a vigorous man spear his own hat with a thrust of his cane. The boys counted four similar hunts in a few minutes.

Women coming around the corner of the Administration Building seemed suddenly impressed with the beauty of the MacMonnies Fountain, and started for it at a run; but, quickly changing their minds, beat back again to their true course. The flags floated stiff upon the gale, and the water in the Lagoon changed color continually.

THE VENETIAN GLASS-BLOWERS.

"I feel," said Mr. Douglass, "that it is rather a pity to spend our last few hours here in a visit to the Midway Plaisance. We should really prefer to go again to the Art Gallery, which we have not half seen."

"I know," said Harry; "but the Art Building is long and time is fleeting. The advantage of going to the Midway is that the poor shows are not worth staying through, and the good ones are few."

Entering the Midway they found that the rain had dampened the enthusiasm of even this crowd—usually the liveliest and gayest on the grounds. They passed by the "Congress of Beauty," and the Philadelphia Working-man's Model Home (Philip wondered why they did not show an Idle-man's Home beside it), and selected the Libbey Glass-works for their first visit.

Within a rounded building they found a tall brick chimney, the lower part of which was made into a glass-heating furnace. About this boys were carrying upon iron rods lumps of what looked like hot coal. When the glass was just right, it was handed to a man, who cut and molded and trimmed the lump of glass into a bottle, or goblet, or globe. The men were so skilful that it was difficult to make out how they did their work; and, somehow, they never seemed to be making any of the more interesting pieces that were exhibited in the show-cases. The only bit of skill the boys could discover was that shown in keeping the ball of molten glass rounded. Whenever one of the rods was put down for a few moments, the glass would become stretched by its own weight to a long drop, and then had to be reheated.

When one of the workmen wished to cut a finished piece of work from the end of the rod, he would hold a pair of cold pincers against it for a few moments and it would snap away at a touch.

The tickets of admission to this show were announced to be good for twenty-five cents applied upon any purchase made in the building. But the boys concluded, after an examination of the prices, that it was easy to see through *that* little scheme. In fact, Harry declared that if postage-stamps had been on sale there, the price of two-centers would have been "two cents and a ticket."

LITTLE DAHOMEY BOY,
AND HIS PLAYTHINGS.

They bought little. Philip paid ten cents "and a ticket" for a spun-glass book-mark, and Harry bought a tiny cup of white and ruby glass. The compartments about the central hall contained, besides show-cases, a loom for weaving glass threads, a glass-cutting wheel, and, most interesting of all, a glass-spinning wheel. The boys studied this for quite a while. There was a big wheel with a broad, thin metal rim kept cool and moist. The workman sat at one side holding a glass rod before a blowpipe and moving it round and round and slowly forward so as to keep it melted fast enough to feed the single long thread to the rapidly revolving wheel.

"How do you suppose he begins the spinning?" said Philip, turning to Mr. Douglass.

"I can only guess," Mr. Douglass replied; "but I suppose he heats a glass rod in the middle, drawing the two ends apart until he makes a thread, and then attaches an end of that thread to the wheel, turning it slowly at first."

"I should think it would be hard to feed the wheel just fast enough," said Harry; "but the man seems to take it easy"; and he did, for he was laughing and winking at the crowd.

AN ACTOR IN THE CHINESE THEATER.

The Venetian Glass-works were just opposite, and as the charge was only ten cents, the party went there also. The process was much the same; but the men were foreigners, and therefore seemed more picturesque. Their work was more interesting to watch. One man was making a sort of spray of glass, and affixed leaves, pressed them with molding-pincers, and twisted them so quickly that it needed close watching for the boys to comprehend the work. He cut the softened glass into scallops with scissors as easily as if it had been dough—every now and then reheating the bit of work. The boys were amused to see him fasten on several ornamental medallions—for he used lumps of red-hot glass for glue.

In the rooms where the Venetian glass was on sale, there was no trace of the businesslike sharpness so noticeable in the American establishment over the way. Here the salesmen moved around as slowly as their own gondolas in contrast with the electric-launch movements of the American shop-people. Leaving the glass-works, they were attracted by a "Japanese Bazaar," and walked through what proved to be only a magnified Japanese store, such as they had often seen. But as they went out, they saw a small boy who was delighted to have found a great cloth fish upon the little lawn outside. With a joyful cheer, he tried to raise it up so that the wind would fill it. But another and very fierce small boy yelled out, "Here, you!—let that fish alone!" and the first boy's cheering stopped at once.

A CHINESE MAMA AND HER BABY.

Upon the same side was the Javanese village, to which they now made a second visit; but it was swept by gusts of cold wind and rain, and bore little resemblance to the sunny, bright little settlement they remembered. The band was silent, there was no chiming of gongs, and the merry little Javanese were soaked and sad. The bazaars, or shop-counters, were deserted except by those on duty, and they were huddled together trying to cheer one another by feeble old Javanese jokes.

Upon the veranda of one of the houses, the boys saw a family of natives at dinner, and one little boy put his hand into the dishes and helped himself. He was not reproved, however, by his father or mother, for they were doing the same thing.

"Here," said Harry, as they passed the middle of the grounds, "is something that only sings louder when it rains"; and he pointed to the musical waterwheel that has been already described; but this time they noticed there was an idol near by—a queer, grotesque figure with which no self-respecting scarecrow would care to claim an acquaintance. He looked as if a hairbrush would have been a shock to his nerves. Only one more thing needs mention.

"There goes *five*," said Philip, and Harry caught sight of a Javanese boy chasing his fleeing straw hat across the road.

Having been advised to see the German village, they took that next, and found it well worth a visit. It contained specimens of old German houses—for instance, a "Black Forest House" dated 1480. The boys and Mr. Douglass walked into its main room, and were becoming a little sentimental over the antique furniture, pictures, and carvings, when a voice brought them back four hundred years by inquiring: "Vill the shentlemans come see my soufenirs? Here are some fine soufenirs!" They declined to see souvenirs, and became absorbed instead in a towel-rack. The roller upon which the towel hung was supported in the hands of a jolly young peasant's figure, who seemed smilingly to hold it forth. Next to this came a Bavarian dwelling, the outside of which was all the party cared to see, for they suspected that the German village required a longer visit than they had intended to pay. And when they had come to the museum, built in the form of a German castle, they were glad they had not stopped to see the Bavarian and other model houses.

INTERIOR OF THE JAVA THEATER.

The collection in the museum began with suits of armor from the rudest of chain-armor to the ornamented plate of later centuries. Arranged upon the wall were specimens of old arms—halberds, pikes, hooks, maces, lances, swords, daggers—every sort of iron tool which would serve to mince one's fellow-man. Besides the array upon the wall, there were show-cases also containing weapons, as well as knives, forks, spoons, and tools.

Philip was amazed to see how much the tools of the Middle Ages resembled those of to-day. What difference there was, told in favor of the old-time workmen, as they seemed to care far more how their instruments looked, decorating the handles and putting ornamental flourishes on the metal parts. The scissors, forks, and knives also were carved and inlaid with gold or silver. Harry saw one enormous pocket-knife that he would have liked to smuggle out. The handle was some six inches long, and the knife had four blades—one plain, one a saw, one a chopper, and the last a pruning-hook. Probably it had been a Christmas present to the head gardener of some Serene High Mightiness, given in recompense for having rescued one of Their Little High Mightinesses from the horse-pond.

The last room in the German castle was filled with dummy figures dressed in various historic German costumes. They were grouped as if attending a reception, and faced a great figure typifying "Germania" surrounded by warriors in helmets and armor.

THE SOUTH SEA ISLANDERS.

Harry said he wondered that "no one had thought of calling the World's Fair City, 'A City of Dummies,'" for all the nations of the earth had gone into doll-making to furnish it with a resident population—a quiet, orderly, law-abiding race, though not full of intelligence.

Just across from the German village, an enormous placard claimed for a Panorama of the Alps the distinction of the "only medal awarded for an exhibit on the Plaisance." It is needless to say that this captured our visitors. They went in and began the ascent of an inclined passage. It curved spirally round and round until they heartily wished it wouldn't. But a party just ahead of them cried out, "Here we are!" and soon they emerged upon a high platform in the middle of the great Swiss Mountains. Harry said he recognized the Matterhorn, the Clatterhorn, the Spatterhorn, and the Flatterhorn; but the lecturer gave other names than these. The lecturer, with frequent allusions to "when I was there," and one condescending "doubtless some of you have *heard* of Interlaken," conducted a sheep-like crowd of sight-seers along a spiral iron fence that was meant to keep people from escaping till they had been at least twice around.

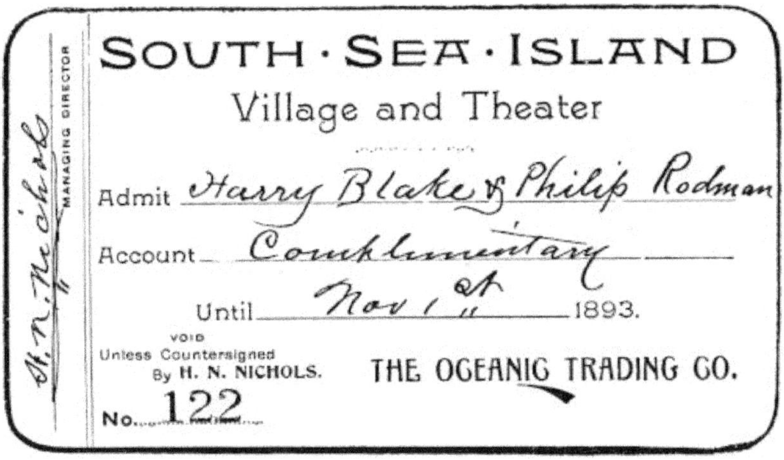

Harry, who was not fond of fences as a rule, took in the situation at a glance, and solved the difficulty by sneaking under the rails to the exit. Philip went after, and Mr. Douglass saw nothing to do but to follow suit. But although they did not care much for the lecturer, the panorama was a fine piece of painting, and Harry said that "if not the Alps it was at least a very good alp for a quarter, even with the lecturer thrown in—still better if he had been thrown out!" But Harry was unfair to this worthy man: most of the visitors enjoyed his clear explanations of the painting, and walked at his heels around all the spirals.

"Samoa"—the South Sea Island show—consisted of a theater and grounds. The grounds were what Philip called "muddish" (a new word to Mr. Douglass, but one he could not disapprove), and the boys stopped only long enough to buy two bark hats,—pointed nightcaps, very elastic and a beautiful brown in color,—and to look in at a Samoan house where, according to the sign, "the boy, for a trifling fee, will show how to kindle a fire by rubbing two sticks together." But the boy sat huddled in a corner, looking as sour as a lemon, and they left him to dream of his native land.

Besides, there was a stamping and a pounding and a yelling going on in the theater that no healthy American boy could long keep away from. When Harry was at the Plaisance one afternoon he had met the manager of this show, and that gentleman had given him passes for the two cousins; so in they went, to find a little stage whereon a gang of savages, naked to the waist, were trying to give people their money's worth so far as stamping, yelling, and racket would avail. They had not even "kept their shirts on," but were all in chocolate-tinted negligé. When the curtain (painted with a Moorish landscape) hid the row, there was only a short intermission before the stage-manager hung out a sign-board announcing a "Religious Dance."

THE ALGERIAN THEATER.

After that act was stilled, Harry said: "I didn't know shinny was a religious dance, but I think now it must be. Perhaps among some of the Pacific Islands a foot-ball scrimmage would be considered a kind of prayer." The dance really was clever, consisting of wheeling about and clattering long and short sticks together rhythmically. The next act was some guttural singing by several women and all the men, who sat in rows cross-legged along the stage. It was just like the song "Swee-ee-eepo—sweepo-o-o! Sweep-ee-o—sweep-o!" that may be heard from certain dusky residents of Manhattan Island on the Atlantic coast. A Fiji war-dance came next, and consisted in showing how bravely they would jab an advancing enemy with a paddle provided he would not go and spoil the little game by warding off or hitting back. It was grand, and the boys were especially delighted to see one of the younger girls come in at the back of the stage and go through the whole dance. "She's a regular Tomboy Fiji," said Philip.

There was more to the program, but the boys tired of it, and betook themselves to one of the Irish villages.

Here were souvenirs of peat, of bog-oak, of lace, all sorts; all sold by tidy little Irish girls with a brogue that it was hard to resist. Mr. Douglass picked up a black bog-oak cane. He seldom carried a cane, and had little idea of buying it. But the Irish girl looked at him with so cordial a smile that he felt bound to say something.

"How much is it?" he asked.

"'Tis a dollar and a half, sir," she answered in a tone of heartfelt regret. Then confidentially, "But it is a fi-i-ine cane, that is, sir!"

He bought it, and the boys grinned. They had seen that the "blarney stone" was at the Fair, and were on their guard. Nevertheless they each bought a tiny black pig cut out of the same bog-oak, and were, as a matter of course, blarneyed in turn.

"How different the people in here!" said Mr. Douglass. "Did you notice that there was a row at the gate, and nothing but joking within?"

"Yes; it reminded me of New York at once," said Harry; "just as Irish Day did."

It was now time for them to leave the grounds; and although they were glad to get home and rest from sight-seeing, they felt very sentimental about taking their last look. They stood in the Court of Honor gazing silently about them, feeling as one feels in giving a parting hand-shake to a loved friend; and then they turned away, knowing that the beautiful dream they had seen and lived in was no more than a dream: that the day would come when all that beauty would be a memory, and the "Ghost City" only a legend.

But the phantom city has taught the American nation that they are a great people, who will some day make true in marble all that was imagined in that short-lived fairy-story of staff.

ONE OF THE TWO IRISH VILLAGES.

CHAPTER XVIII

Packing for Home — A Glimpse of Niagara — Philip tells his Adventure — Foiling a Clever Swindler — A Convincing Exposure.

THEY packed up that same afternoon, after considerable trouble in finding room for the knickknacks they had picked out, and took a carriage to the station. They found no signs to direct them, and had to inquire several times to make sure which was the track upon which they might expect the train for New York. When they thought they were certain of this, they saw a train come in on schedule time and on the proper track. But, to their surprise, a man called out, "Illinois Central train for Cairo and St. Louis!" which threw the crowd into an uncertainty anything but pleasant. Missing a local suburban train is troublesome enough; but missing a train that is going one third across a continent is a disaster.

To their great relief, the obliging colored porters very promptly corrected the stupid error, and they found themselves safely upon the train for New York.

Their journey was a repetition of their trip out, except that this time they stopped to see the Falls of Niagara, viewing them, as young Phinney had done, from above the falls.

"It's a pity to see them from a distance only," said Mr. Douglass. "I should like to stay awhile."

"They are well worth going over carefully," said Harry, thoughtfully; and Philip looked at him inquiringly.

During the second day on the train, Mr. Douglass was talking to the boys as to their experiences at the Fair; and then Philip's little adventure, before referred to, came out. As he told the story it ran something like this:

THAT day when I was taking photographs in the Plaisance, I went into the Cairo Street a second time. I wanted, if possible, to get a picture of the little boy who leads the camels. They stopped me at the door, and while explaining that I had been permitted to take photographs there, I put my camera for a few moments on a camp-chair.

A KODAKER CAUGHT.

When I looked around for it, my camera was missing. I tell you, I felt pretty mean. At first I didn't know what to do. I asked the ticket-taker about it, but he hadn't seen any one take it. Then I thought, quick, what a man would do who had picked up a camera like that, and I made up my mind that he would want to get out of Cairo Street as fast as he could. Of course, most of the people there were sight-seeing, and just moved along slowly. So I hopped up on top of the camp-chair, and looked over the crowd. Luckily, I caught sight of a man with a brown felt hat, who was moving fast through the slow-moving people. I made up my mind that it was my last chance for my kodak, and I went through the crowd like a snowplow through a drift. I kept my eye on that brown felt hat, and pretty soon I caught up to the man. Once I thought I had lost him, for a camel came by, and I had to get out of the way; but I found him again, and, as I said, I got near to him.

I saw at once that he had a camera in his hand, and I was pretty sure it was mine. But just as I was going to catch hold of it, I happened to think it was a serious matter to tell a man he was a thief, and I stopped to make sure what I ought to do. The man was pushing through the crowd so fast that I had no good chance to take a real square look at the camera, so I concluded I would just keep after him till he thought he was clear away. He kept looking behind him at first, but now he began to go slower, as if he thought everything was all right. ["Little dreaming," Harry put in, "that a sleuth-hound wearing magnifying-glasses was upon his tr-rail!"]

I kept off to his left, and he didn't see me. Pretty soon he came out into the Fair Grounds, and there weren't so many people there. He turned toward one of the north entrances, and I kept a sharp lookout for a Columbian Guard. I didn't take the first one I saw, because he looked sleepy and stupid, and I was afraid he would arrest *me*; but the next was a soldierly-looking fellow, and after seeing my man was taking it easy, I went to this guard and said:

"That man with the brown felt hat, there, picked up my camera when I wasn't looking, and walked off with it. I want you to get it back for me."

"Sure, young fellow?" he said, looking at me hard.

"Sure," I said; for by that time I had seen a bruise on one corner of the camera where I dropped it once.

"All right," said he. "Come along. You go after the man, and don't lose sight of him, and I'll go around this little building and meet him."

So we did. And it worked first-rate. The guard was a fly sort of a fellow, and instead of asking the man whether that was his camera, he asked him whether he had a permit for it.

The man stopped and looked puzzled for a minute, then he put on a face as bold as brass, and said: "No, sir. I have not yet obtained one, but I was going to get one."

"Where were you going for it?" said the guard, to catch him.

REGISTERING IN NEW YORK STATE BUILDING.

"I was about to ask you," said the man, with a sharp kind of a smile, seeing the guard's little game. This made the guard lose his temper, and out he came with the whole story.

"This young man here says that camera is his, and that you picked it up," said the guard.

"The impertinent young rascal!" said the man, who must have been a cool hand.

"We'll see about that," said the guard, who began to wonder which of us was lying.

"I don't propose to be bothered by this young scamp," said the man, seeing that the guard hesitated a little. "If you will tell me where to obtain a permit for my camera, I shall be obliged to you."

Well, his coolness staggered the guard, and it did me. I wondered for a minute whether I had made a mistake; but when I looked at the camera, there was the bump on the corner, and I was sure again.

"Ask him," I said to the guard, "what is the name of his camera."

"You saucy young villain, I don't propose to be questioned about this any longer!" said the man, and he turned to walk away. But that decided the guard.

"No, sir!" he said. "You'll come with me, and we'll have this question settled."

The man looked around quick, as if he was wondering what the chances were if he should run for it; but the guard laid his hand on the man's shoulder, and the swindler then decided to brazen it out.

"Very good," he said, looking at his watch; "I shall lose my train, but I suppose this absurd matter must be disposed of."

"But I thought you wanted a permit for your camera?" said the guard, with a grin; and then the man bit his lip. That time *he* made a mistake.

The guard went to a sort of little sentry-box, and sent out a signal. Pretty soon a patrol-wagon came driving up, and we were taken in it outside of the grounds to a police-station.

"Officer," said the man to the sergeant (I suppose it was), "this foolish boy has laid claim to my camera, and—"

ALONG THE LAKE.

"Now, don't be in a hurry," said the officer, coolly. "I'll hear the guard first, please." The guard told the story very clearly and plainly.

"Is that correct?" said the sergeant to me.

"Yes, sir; and I can prove—" I began.

"Go slow, young man," said the sergeant, motioning to me to stop talking. Then he said to the man who had my camera:

"Is that story correct?"

"Entirely, Sergeant."

"Very good," the sergeant said. "Now, young man, how can you prove it is your box?"

"Well," said I, "it's a Kodak No. 4, and it has a bruise on one corner."

"Yes," said the swindler, "I see. That is what has caused the trouble. Mine has a bruise on the corner, too. I dropped it this morning as I was coming through the turnstile."

"That's rather slim proof to arrest a man on," said the sergeant, looking hard at me. Then I began wondering how I could prove my ownership, and I thought of the pictures I had taken.

"I know!" I said. "I can prove it by the photographs I took. I remember some of them anyway. There was one of—"

THE DARK ROOM.

"Hold on!—hold on!" cried the sergeant, quick as lightning. "It's the defendant's turn now. Perhaps, sir, you will tell us what pictures are in the camera?"

"I am sorry to say that I cannot," said the man, still polite. He was a smart fellow. "Indeed, the camera belongs to a friend of mine, and he lent it to me this morning for the day. He may have taken pictures with it. I took only one myself, and that was a view of the crowd in Cairo Street. If you will have the pictures developed, you will see that I am right."

Then I was scared. I wish you could have seen the fellow—he was as cool as a cucumber. He was no common swindler, I'm sure.

"That's a fair proposal," said the sergeant, who was puzzled by this queer case. "Let us adjourn to a photographer. And don't let either of these men get away," he added, turning to a policeman.

So then we formed in procession, and went around the corner to a photographer's and into his dark-room. The sergeant explained what we wanted.

But before the photographer began to develop the film, I spoke up and said: "Sergeant, this man probably took one picture just after he picked up the camera. It was all set, and all he had to do was to touch the button. Now, it isn't likely he knows anything about the camera if he stole it. If he didn't, his friend must have told him how to work it."

"I think that's a sound argument," said the sergeant. "But suppose you write down all the pictures you remember taking."

"I don't know how to manage the camera entirely," said the man; "but I intended to get the photographer to explain it to me."

While I was writing down all the pictures I remembered, and the photographer was developing the film, the sergeant turned to the man who had taken my camera, and said quickly: "By the way, what was the name and address of the friend who lent you the camera?"

Well, that staggered the fellow completely. "I brought it from New York," he began, "and his name is—"

LUNCHING OUTDOORS.

"Don't trouble yourself to invent a name," said the sergeant, sharply. "You said he lent it to you this morning for the day. Now, I doubt whether you came from New York this morning. Don't you think that you *may* have picked up this camera by mistake for the one your friend lent to you this morning in New York?"

But before the fellow could answer, the photographer said: "The pictures tally with the young man's list, and the one of the crowd in Cairo Street is a double exposure showing that the film hadn't been wound up after this young man had taken the previous picture outside.

"And, Sergeant, the funniest part of it all is, that one of the pictures that the young man took just at the door of Cairo Street, shows this man standing looking at the camera, but without any of his own!" and then all the men in the room looked at the thief and grinned.

"Well," said the sergeant to the man, "what do you think about that mistake?"

"I'm afraid it must have been an error," said the man, rather shakily. "I picked up this camera thinking it was my own, and—"

The sergeant said sharply, "Now, you get out of here, and quick too. It wouldn't pay to prosecute you, for you're too slippery. Get out—quick!" And the man just skipped.

"Now, young man," said the sergeant, "you take better care of your camera next time. I'll see you into the grounds again."

So I thanked him. He saw me through the gate, and that was the end of my adventure. But it was a close shave. I didn't tell you about it before, for fear you would think I had been stupid.

WONDERFUL!

THE FERRIS WHEEL, FROM "OLD VIENNA."

CHAPTER XIX

Mr. Douglass has a Remarkable Experience.

"NO, sir; not this afternoon, sir. I'm very sorry, but that's the orders. We have to be very careful with her, sir. There hasn't been anybody in it for full two hours," said the man at the gate.

"But it's one of the advertised attractions of the Midway, and I insist," said Mr. Douglass. He had already been in the Ferris Wheel once before, and had not meant to return to it, but circumstances were too strong for him, and here he was, ready to pay, but unable to get a ticket.

"Insist or not," said the man at the gate; "you can't get in if you want to; we can't let you in if we want to. The wheel is sulky, and has been turning slow and ugly like that since noon to-day."

"But I leave the city to-night," said the tutor, "and I will not leave without another ride in the great wheel."

"Very good," said the man, turning on his heel; "get in if you can. The machinery is out of order, and we can't stop the wheel—maybe you can"; and he walked off whistling "Comrades."

The man's indifference roused Mr. Douglass. "We'll see," said he, "whether I won't have one more ride on the Ferris Wheel!"

After a brief glance around him, his eye caught the sign of the Bedouin encampment. Rushing toward it, he threw a twenty-dollar gold piece upon the counter, told the attendant to keep the change, and was soon in earnest conference with the Arab sheiks.

He gave each a golden double eagle, and they bowed low. "Allah be praised, the white chief's will shall be done!" they exclaimed.

Then, without losing a moment, the three hurried to the great Ferris Wheel, which still went painfully, jerkily about, with a low growl that boded mischief. But if the wheel was out of temper, so was Mr. Douglass; and, saying "Ready!" to the Arabs, he placed himself between them, one grasping each of his arms. "Let go!" the tutor called; and at the word, the sinewy Arabs raised him from the ground, and, after one or two preliminary swings, hurled him through the air as if he had been a stone from a sling.

A GLIMPSE OF THE HORTICULTURAL DOME.

Crash! went the tutor through the glass, just scraping his way between two of the iron bars, but landing safely in a car.

"There!" he cried, "I *shall* have another ride in the wheel!"

Up it went, over, down, and he came slowly toward where the Arabs stood in earnest talk. As he approached, one stepped forward:

"Give more bakshish!" he cried, "or—"

Mr. Douglass shook his head. The Arabs shook their fists. He laughed at them. Then, raging with fury, one turned and said in Arabic to the other:

"Seeme letim sleyd!"

No sooner said than done. Each Bedouin seized one of the gigantic supports that upheld the wheel, and pulled with all his might. They were both well-developed and had a strong pull. With a long pull and a strong pull and a pull all together, they sprung out the supports, the great wheel fell from its place, and the Bedouins, seeing the mischief they had done—and perhaps repenting of it, for they were only hasty, not wicked—leaped upon their priceless donkeys, and were soon lost in the suburbs of Chicago. Unlike the cat, they did not return, and have nothing more to do with the story. But no doubt they often regretted, as they grew older, the hasty outburst of temper that was now to do so much mischief.

For the wheel, with Mr. Douglass an unwilling passenger, dropped to the ground, and rolled slowly up the Plaisance.

Its first victim was the Turkish village, and when the wheel had passed, the village looked like a flat, hand-colored map.

THE FISHERIES BUILDING, FROM ACROSS THE LAGOON.

Mr. Douglass, as soon as he saw what the Arabs were at, had climbed out of the car, and, more like a spider than a tutor, made his way to the axle, where he stood upright, walking backward upon the axle as the wheel ran forward. From this well-chosen perch, he could, and did, witness the ensuing scene—which was described by the Chicago reporters as "unusual."

The Turkish village, being a trifle lumpy, diverted the wheel but little, and the next assault was upon the corner of the Panorama of the Alps. The end of the canvas became entangled in the wheel, and was stretched from one side to the other, so that subsequently many thought that there had been a land-slide when they saw the wheel pass.

Mr. Hagenbeck's far-famed Animal Show also came in for a share of damage, the wheel crushing one corner of the menagerie, and picking up the small performing-bear in such a way that he was compelled to leap from car to car as each came upright, and walk the wheel as if it were a circus ball. He was rescued unhurt, but considerably fatigued, when the wheel finally—but it was not yet through.

Glancing to the other side of the Plaisance, the Libbey Glass Company was splintered into what one of the Irish dairymaids declared to be "smithereens," and the monster rolled onward to where the International Dress and Costume Exhibit was situated. Here it broke in one side of the building, and then, catching sight of the contents, with a shriek from every cog fled into the Fair Grounds, cutting its way through the Illinois Central and Intramural bridges, with no more than slight crunches. The bear and Mr. Douglass were still walking their tread-mills, and the Panorama of the Alps still decorated a whole side of the wheel.

AT A DRINKING-FOUNTAIN.

But the great wheel, though out of temper, was not yet without feeling. It swerved aside upon reaching the Woman's Building, plunged into the Lagoon, where, frightened by the squawking of the swans, it shot madly toward the Government Building. Probably it would have gone entirely through except for the fact that the Department of Justice lay directly in its course. It could not face the stern portraits of judges upon its walls, and, destroying only the big tree and a few other antiquities of slight importance, it encountered the Liberal Arts Building but slightly checked in speed.

Mr. Douglass was tired of his ride, and, from the bear's growling, concluded that his fellow-passenger was also ready to stop.

"I wish," said Mr. Douglass (never relaxing his backward walk), "that I had omitted this last visit to the Fair. It is rather exciting, but too wearisome after my long weeks of tramping. I am glad to see the Building of Manufactures ahead. The wheel may get through it, though I couldn't; but it won't go much farther."

But he was wrong. The lath and plaster offered little resistance to the iron wheel, and the little elevator boy in the center of the building opened all the throttles, shot bodily out through the roof, elevator and all, and landed in the Viking ship, much put out but little hurt.

A LITTLE VISITOR

On its way down the center aisle, the wheel picked up the big telescope, and on its next revolution flung that marvelous instrument high in air. But Ben Franklin was waiting for just such a chance, and he promptly accepted it. Chucking aside his key and kite-line, he stepped lightly out from the portico of the Electricity Building and caught the telescope on the fly (for which feat he afterward received a vote of thanks from the University of Chicago), placed it carefully on the Wooded Island, and modestly resumed his place on the pedestal, saying simply, "A penny saved is two pence clear."

Cutting a clear channel through the biggest building, the wheel leaped the Basin—a sight that so astonished Miss Progress that she called to the Sciences and Arts to save themselves, came down from her perch, fled shrieking into Machinery Hall, and took the Crane for the other end.

Miss Republic noticed the passing of the wheel, but, until it was gone, did not understand what was going on.

The wheel was now headed directly for Agricultural Hall, but as it came within a threatening distance, the three young women of the Zodiac family, with a single impulse, threw their globe at the wheel—at the same time uttering three shrieks that did more execution than the ball they had thrown. The ball shattered one of the towers on the Convent; but the shrieks saved the Agricultural Building, with all its priceless corn-cobs, preserved prunes, and patent harvesters.

Scared from its course, the wheel sought an avenue of escape. To Mr. Douglass's horror, and the bear's regret, its course lay toward the Moving Sidewalk. Striking the wrong (the incoming) side, the wheel began to see that it had made a mistake, for gradually it was compelled to slow up.

Mr. Douglass and the performing-bear seized the opportunity to take a short rest. Both were experienced travelers, and never failed to take advantage of any chance to relieve the monotony of a journey.

Meanwhile, an alarm had been sounded upon all the trumpets held by figures upon the Administration Building; telephones were at work calling aid; the Fire Queen and all the patrol-wagons were dashing to and fro; the Krupp gun was loaded and trained upon the wheel; and all was bustle and excitement.

Buffalo Bill, Texas Jack, and Professor Hagenbeck with high boots on, came riding like mad across the Court of Honor, and charged bravely down upon the motionless wheel. When within range, Mr. Cody opened fire, and succeeded in breaking all the windows that still remained intact in any of the wheel cars. Texas Jack lassoed the bear, and dragged the grateful beast from the top of the wheel, whereupon the professor consoled the little animal by giving him the usual lump of sugar taken from the professor's coat-tail pocket. Just at this moment, Engines "999," "John Bull," and "De Witt Clinton" arrived for the purpose of hauling the wheel back to its place.

They were just too late.

The wheel having lost headway and remained still for a short time, now began to be carried back along the sidewalk. It rounded the curve, ran along the pier to the end, and, on coming back, had acquired a speed that sent it off upon a new expedition.

This time the Statue of the Republic realized there was something irregular in the action of the wheel, and aroused from her lethargy enough to step languidly ashore and let the wheel go by. The Krupp gun was discharged, but the missile, missing the wheel, put an end to the battle-ship "Illinois," who went into plaster chips with her flag still flying.

Mr. Douglass said, pettishly, "I am getting very much bored at having to run about on this axle, and I do think the authorities of the Fair ought to do something to protect a visitor from such an accident."

But his conscience told him that he had done wrong in entering the wheel without having secured permission.

As the great unicycle ran for the Transportation Building, the statue of "Land" remarked, "For the land's sake!" and hastily put on brakes, a course for which he was commended by Messrs. Fulton and Watt, his neighbors. Stephenson, however, blamed him for not first securing one of the air-brakes, of which there were plenty inside the building.

Striking the Intramural line, the wheel ran over Festival Hall, exploding the bellows of the great organ, and then ran triumphantly up and bursted the Horticultural bubble of glass.

Just here, however, the wheel and Mr. Douglass caught sight of the dome of the Illinois State Building, and the iron creature turned aside with a sigh that could be plainly heard at the British Building on the lake-shore, and then ran down the Midway like a hunted stag.

Here Professor Hagenbeck and his young men received the wheel with stern glances that even that awful monster of iron and glass found irresistible. With a few lashes of his long whip, the professor soon reduced the wheel to submission, and at the word of command it ran to its place, climbed into position, and was still. The professor immediately gave the wheel a lump of sugar from his coat-tail pocket, patted it upon the cogs, and saying, "There will be no further trouble, I think," walked serenely back to lunch.

He had forgotten Mr. Douglass!

How was the poor tutor to reach the ground?

He tried to climb down one of the spokes, but slipped, lost his hold, and was falling, falling, fall—

"I really believe, boys," said Mr. Douglass, "that I've been asleep. I've had a remarkable dream. It was—" But the brakeman called:

"New York, last stop, all out!"

THE 194,000,000 CANDLE-POWER SEARCH LIGHT.

IN THE MIDWAY PLAISANCE.

www.ingramcontent.com/pod-product-compliance
Ingram Content Group UK Ltd.
Pitfield, Milton Keynes, MK11 3LW, UK
UKHW040815280325
456847UK00003B/420